Praise for Tali Sharot's

The Optimism Bias

"Very enjoyable, highly original and packed with eye-opening insight, this is a beautifully written book that really brings psychology alive." —Simon Baron-Cohen, author of *The Science of Evil*

"Offers evolutionary, neurological, and even slightly philosophical reasons for optimism. . . . A book I'd suggest to anyone." —Terry Waghorn, *Forbes*

"If you read her story, you'll get a better grip on how we function in it. I'm optimistic about that." —Richard Stengel, *Time*

"Once I started reading *The Optimism Bias*, I could not put it down." —Louisa Jewell, *Positive Psychology News Daily*

"An intelligently written look into why most people take an optimistic view of life. . . . [A] fascinating trip into why we prefer to remain hopeful about our future and ourselves." —*New York Journal of Books*

"With rare talent Sharot takes us on an unforgettable tour of the hopes, traps, and tricks of our brains. . . . A must-read." —David Eagleman, author of *Incognito*

"A fascinating yet accessible exploration of how and why our brains construct a positive outlook on life."
—BrainPickings.org

"Lively, conversational. . . . A well-told, heartening report from neuroscience's front lines." —*Kirkus Reviews*

"Most readers will turn to the last page not only buoyed by hope but also aware of the sources and benefits of that hope." —*Booklist*

"Fascinating and fun to read. . . . Provides lucid accounts of [Sharot's] often ingenious experiments."
—*BBC Focus Magazine*

Tali Sharot

The Optimism Bias

Tali Sharot's research on optimism, memory, and emotion has been the subject of features in *Newsweek*, *The Boston Globe*, *Time*, *The Wall Street Journal*, *New Scientist*, and *The Washington Post*, as well as on the BBC. She has a Ph.D. in psychology and neuroscience from New York University and is currently a faculty member of the Department of Cognitive, Perceptual, and Brain Sciences at University College London. She lives in London.

www.theoptimismbias.com

The Optimism Bias

The Optimism Bias

A Tour of the Irrationally Positive Brain

Tali Sharot

Vintage Books
A Division of Random House, Inc.
New York

FIRST VINTAGE BOOKS EDITION, JUNE 2012

Portions of this work were previously published in different form in *TIME Magazine* (May 2011).

The Library of Congress has cataloged the Pantheon edition as follows:
Sharot, Tali.
The optimism bias : a tour of the irrationally positive brain / Tali Sharot.
p. cm.
Includes bibliographical references and index.
1. Neuropsychology. 2. Optimism—Physiological aspects. I. Title.
QP360.S466 2011
612.8—dc22 2010039692

Vintage ISBN: 978-0-307-47351-6

Book design by M. Kristen Bearse

www.vintagebooks.com

Printed in the United States of America
10 9 8 7 6 5 4 3 2 1

For my parents—

Tamar and Steve Sharot

Contents

Prologue

A Glass Forever Half Full?

I would have liked to tell you that my work on optimism grew out of a keen interest in the positive side of human nature. That would be a pleasant story: "A Cognitive Neuroscientist in Search of the Biological Basis of Our Hopeful Souls." Pleasant but, unfortunately, untrue. I stumbled upon the *optimism bias* quite by accident while investigating people's memories of the largest terrorist attack of our time. Back then, my scientific interests inclined more to the dark side: My principal research had been aimed at understanding how traumatic events shape our memories. I was interested in how the brain tricks us into believing that our recollections of exceptionally emotional events, such as the occurrences of September 11, 2001, are as accurate as a videotape, even when we are utterly mistaken.

I had been conducting research at New York University for over a year when American Airlines Flight 11 and United Flight 175 were flown into the World Trade Center at 430 miles per hour. Shock, confusion, and fear were the common responses on the street. Such forceful emotions are exactly the sort of reactions that will generate unusually vivid memories, ones that are reluctant to fade away. These are commonly referred to as "flashbulb memories" because of their sharp-edged, picturelike qualities. In chapter 9, I tell the story of flashbulb memories—how we remember unexpected arousing events and how the structures deep in our brain "Photoshop" these images, adding contrast, enhancing resolution, inserting and deleting details.

I was puzzled: Why had our brains developed a mechanism that would create highly vivid memories that were not necessarily accurate? Around the time my colleagues and I published our scientific investigation of memories of 9/11,[1] a group of researchers at Harvard University proposed an intriguing answer. The neural system responsible for recollecting episodes from our past might not have been developed for that purpose at all. Rather, the core function of this system, which many had believed evolved for memory, may, in fact, be *to imagine the future.*[2]

Brain-imaging studies show that the same brain structures that are engaged when we recollect our past are called upon when we think of the future.[3] These two fundamental human thought activities rely on the same brain mechanisms; they draw on similar information and underlying processes. To imagine your upcoming trip to Barbados, for example, you need a system that can flexibly reconstruct novel scenarios, one that can take bits and pieces of memories from your past (your last vacation to a warm country, images of sandy beaches, your partner in his swimsuit) and bind them together to create something new (you and your loved one wearing straw hats on a beach in Barbados next month)—an event that has yet to happen. Because we use the same neural system to recall the past as we do to imagine the future, recollection also ends up being a reconstructive process rather than a videolike replay of past events, and thus is susceptible to inaccuracies.

Was this theory correct? To discover the answer, I would record people's brain activity while they imagined *future* events, then compare that activity to the pattern I observed when they were recollecting *past* events.

The plan was simple. However, when I asked my volunteers

to imagine future life events, something unexpected occurred. Even when given specific situations of the most humdrum kind (getting an ID card, playing a board game), people tended to fashion magnificent scenarios around them. They kept painting the most perfectly gray events in shades of pink.

You would think that imagining a haircut in the future would be somewhat dull. Not at all. Getting a haircut today may be boring, but getting one in the future is a cause for celebration. Here is what one of my participants wrote:

> I projected that I was getting my hair cut to donate to Locks of Love [the nonprofit organization that provides hairpieces to children suffering from hair loss]. It had taken me years to grow it out and my friends were all there to help celebrate. We went to my favorite hair place in Brooklyn and then went to lunch at our favorite restaurant.

I asked another participant to imagine a ferry ride. She responded:

> One to two years from now I see myself taking the ferry to the Statue of Liberty. The weather would be really nice and windy, so my hair would be blowing everywhere.

The world, only a year or two into the future, was a wonderful place to live in. I spent hours with a student of mine, Alison Riccardi, trying to come up with exceptionally unexciting events that surely couldn't provide any cause for celebration. All to no avail. Once people started imagining, the most banal life events seemed to take a dramatic turn for the better, resulting in a life that was just a bit less ordinary.

These responses switched on a red (or at least pinkish) light in my mind. I was surprised by this tremendously powerful, seemingly automatic inclination to imagine a bright future. If all our participants insisted on thinking positively when it came to what lay in store for them personally, then there had to be a neurobiological basis for this phenomenon. We set aside our original project and went on to try to identify the neural mechanisms that mediate our optimistic tendencies.[4]

How does the brain generate hope? How does it trick us into moving forward? What happens when it fails? How do the brains of optimists differ from those of pessimists? Although optimism is vital for our well-being and has an enormous impact on the economy, these questions have been left unanswered for decades. In this book, I argue that humans do not hold a positivity bias on account of having read too many self-help books. Rather, optimism may be so essential to our survival that it is hardwired into our most complex organ, the brain.

From modern-day financial analysts to world leaders, newlyweds (all described in chapter 11), the Los Angeles Lakers (chapter 3), and even birds (chapter 2), optimism biases human and nonhuman thought. It takes rational reasoning hostage, directing our expectations toward a better outcome without sufficient evidence to support such a conclusion.

Close your eyes for a moment and imagine your life five years from now. What sorts of scenarios and images pop into your mind? How do you see yourself professionally? What is the quality of your personal life and relationships? Though each of us may define *happiness* in a different way, it remains the case that we are inclined to see ourselves moving happily toward professional success, fulfilling relationships, financial security, and stable health. Unemployment, divorce, debt, Alzheimer's, and any number of other regrettably common misfortunes are rarely factored into our projections.

Are such unrealistic predictions of future bliss limited to thoughts of fundamental, life-altering events such as marriage and promotion? Or do optimistic illusions extend to more mundane, everyday events? Do we expect to get more work done this week than last? Do we expect tomorrow to be better than yesterday? Do we assume that next month will be filled, on the whole, with more enjoyable encounters than irritating ones?

In the summer of 2006, I set out to research this more prosaic part of the equation. I was spending a few months working at the Weizmann Institute for Science in Israel before starting a new job at University College London. Whatever the depth of my own optimistic nature, I wasn't banking on getting too many sunny days once I got to the United Kingdom, so I was fairly determined to get some sun on my face for a few weeks before relocating to the English capital.

The Weizmann Institute is about a twenty-minute drive from the bustling city of Tel Aviv. It's a scientific oasis in the middle of the country, its well-tended greens reminiscent of those on a California campus. But if the institute feels peaceful in itself, it is no secret that the volatile politics of Israel are always close at hand. Most Weizmann students enter college after fulfilling their mandatory military service, a life experience that doesn't necessarily incline one to be optimistic. With this in mind, I wondered about the extent to which they would be prone to the optimism bias. I recruited my sample group, then asked them about their expectations for the month ahead. On the most humdrum level, how likely did they think it was that they would be stuck in traffic sometime, or be more than half an hour late for an appointment? On a slightly more heightened level of anticipation, what did they think were their odds of having a sexual encounter they would regret, or one they would enjoy? Did they see themselves cooking an elaborate meal, receiving a surprise gift? I presented them with one hundred such questions.

I have to say, the results astonished me, in that an overwhelming majority of the students expected more positive experiences than negative or even neutral ones, at a comparative rate of around 50 percent to 33 percent. That was not all: The positive incidents were also expected to occur sooner than the unpleasant or just plain boring ones. While the students generally expected to enjoy a nice evening out with a partner in the next few days, an argument with a boyfriend or girlfriend was anticipated—if at all—only toward the end of the month.

On the off chance that my participants were leading charmed lives, I asked them to come back a month later and tell me which of those one hundred notional events they had actually experienced during that time. As it happened, positive, negative, and neutral everyday events had befallen them more or less equally, in roughly even 33 percent shares. The Weizmann students had not hit on the secret of human happiness; they had merely expressed a very ordinary optimism bias.

In considering this example, you may be wondering whether optimism is truly a dominant trend in the population at large or, more specifically, a special delusion of youth. This is a fair question. You would think that as we grow older, we grow wiser. With more years of life experience behind us, we should be able to perceive the world more accurately—to distinguish delusions of hope from hard-core reality. We should, but we don't.

We wear rose-tinted glasses whether we are eight or eighty. Schoolchildren as young as nine have been reported to express optimistic expectations about their adult lives,[5] and a survey published in 2005 revealed that older adults (ages sixty to eighty) are just as likely to see the glass half full as middle-aged adults (ages thirty-six to fifty-nine) and young adults (ages eighteen to twenty-five).[6] Optimism is prevalent in every age group, race, and socioeconomic status.[7]

Many of us are not aware of our optimistic tendencies. In fact, the optimism bias is so powerful precisely because, like many other illusions, it is not fully accessible to conscious deliberation. Yet data clearly shows that most people overestimate their prospects for professional achievement; expect their children to be extraordinarily gifted; miscalculate their likely life span (sometimes by twenty years or more); expect to be healthier than the average person and more successful than their peers; hugely underestimate their likelihood of divorce, cancer, and unemployment; and are confident overall that their future lives will be better than those their parents put up with.[8] This is known as the optimism bias—the inclination to overestimate the likelihood of encountering positive events in the future and to underestimate the likelihood of experiencing negative events.[9]

Many people are convinced that optimism was invented by Americans—a by-product of Barack Obama's imagination, some believe. I encounter this notion often, especially when giving lectures in Europe and the Middle East. Yes, they say, celebrating future haircuts, imagining a sun-filled ferry ride, underestimating the likelihood of crippling debt, cancer, and other misfortunes are indicative of an optimism bias—but these are New Yorkers you are describing.

True, my first investigation of optimism was conducted on residents of Manhattan. (I made a special effort to conduct all future investigations on cynical Brits and Israelis.) You could be forgiven for assuming that the Big Apple is the perfect setting for concerted research into optimism. Though I have no hard statistics at hand to substantiate this, pop culture would certainly have us believe that New York City is a lightning rod for individuals with big dreams and the self-belief to think these can be realized. From newly processed immigrants contemplating the Statue of Liberty to Holly Golightly admiring the window displays at Tiffany's on Fifth Avenue, NYC is quite the poster child for all things hopeful: a city of teeming streets where people are continually hustling to be the Next Big Thing.

To the surprise of some, however, the concept of optimism can easily be traced back to seventeenth-century European thought. The formulation of an optimistic philosophy took root not in American culture but in France. Descartes was one of the first philosophers to express optimistic idealization, in his trust that humans could master their own universe and thereby enjoy the fruits of the earth and the maintenance of good health. But the introduction of *optimism* as a technical term is usually credited to the German philosopher Gottfried Wilhelm Leibniz, who notably held that we live in "the best of all possible worlds."[10]

The results of having a positively biased view of the future can be quite dire—bloody battles, economic meltdowns, divorce, and faulty planning (see chapter 11). Yes, the optimism bias can sometimes be destructive. However, as we will soon discover, optimism is also adaptive. As with all other illusions of the human mind (such as the vertigo illusion and the visual illusions described in chapter 1), the optimism illusion had developed for a reason: It has a function.

The optimism bias protects us from accurately perceiving the pain and difficulties the future undoubtedly holds, and it may defend us from viewing our options in life as somewhat limited. As a result, stress and anxiety are reduced, physical and mental health are improved, and the motivation to act and be productive is enhanced. In order to progress, we need to be able to imagine alternative realities—not just any old realities, but better ones, and we need to believe them to be possible.

The mind, I argue, has a tendency to try to transform predictions into reality. The brain is organized in a way that enables optimistic beliefs to change the way we view and interact with the world around us, making optimism a self-fulfilling prophecy. Without optimism, the first space shuttle might never have been launched, peace in the Middle East would never have been attempted, rates of remarriage would likely be nonexistent, and our ancestors might never have ventured far from their tribes and we might all be cave dwellers still, huddled together and dreaming of light and heat.

Fortunately, we are not. This book explores one of the greatest deceptions of which the human mind is capable: the optimism bias. It investigates when this bias is adaptive and when it is destructive, and it provides evidence that moderately optimistic illusions can promote well-being. It focuses on the specific architecture of the brain, which allows unrealistic optimism to be generated and alter our perceptions and actions. In order to understand the optimism bias, we first need to look at how, and why, the brain creates illusions of reality. We need to burst a giant bubble—the notion that we perceive the world as it really is.

The Optimism Bias

CHAPTER 1

Which Way Is Up?

Illusions of the Human Brain

January 3, 2004, Sharm el-Sheikh. One hundred and forty-eight passengers and crew board Flash Airlines Flight 604 bound for Paris via Cairo. The Boeing 737-300 takes off at exactly 4:44 a.m. Two minutes later, it disappears from the radar.

Sharm el-Sheikh is located on the southern tip of the Sinai Peninsula. It is a tourist destination because of its year-round warm weather, beautiful beaches, and marvelous snorkeling and diving. The majority of passengers on Flight 604 are French tourists escaping the European winter to spend their Christmas vacation near the Red Sea. Entire families are on board Flight 604, on their way back home.[1]

The crew is largely Egyptian. The pilot, Khadr Abdullah, is a decorated war hero, because of his performance flying the MiG-21 in the Egyptian air force during the Yom Kippur War. He has 7,444 flying hours under his belt, although only 474 of those are on the Boeing 737 he is piloting on this day.[2]

According to its designated route, the aircraft should have ascended for a short while after takeoff and then turned left, heading toward Cairo. Instead, less than a minute into the flight, the plane turns right and quickly assumes a dangerous angle. Flying completely on its side, the jet begins spiraling downward toward the Red Sea. Just before impact, the pilot appears to gain control over the now upside-down plane, but it is too late.[3] Flight 604 crashes into the water moments after takeoff. There are no survivors.

At first, the authorities suspect a bomb had been planted on the plane by terrorists. This hypothesis arises because no distress signal was sent from the aircraft. However, when the sun comes up and pieces of the jet are discovered, it becomes apparent this theory is wrong. The pieces of the plane are detected close together, and there are not many of them.[4] This suggests that when the plane hit the water, it was intact, rather than having exploded in midair, which would have resulted in many fragments scattering across the sea. What, then, caused Flight 604 to drop violently from the sky?

For the mystery to be solved, it is essential that the plane's black box be found. The area of the sea where the plane crashed is one thousand meters deep, which makes it difficult to detect the signals emitted from the box. Furthermore, the black box's battery will last for only thirty days; after that, the probability of finding it will be, realistically, nil. Egyptian, French, and U.S. search teams participate in the effort. Luckily, two weeks into the search, the black box is detected by a French ship.[5]

The information from both the data recorder and the voice cockpit log contain clues that guide the investigators in a number of different directions. No less than fifty different scenarios are identified, then ruled out one by one, on the basis of the available data. No evidence of any airplane-related malfunction or failure can be found.[6] The investigators are left with a handful of scenarios, which they then try out in a plane simulator. After examining the remaining scenarios thoroughly, all but one are deemed inconsistent with the data at hand. The U.S. research team concludes that "the only scenario identified by the investigative team that explained the accident sequence of events, and was supported by the available evidence, was a scenario indicating that the captain experienced spatial disorientation."[7]

During spatial disorientation, also known as vertigo, a pilot is unable to detect the position of the aircraft relative to the ground. This usually happens when no visual cues are available, such as when the plane is flying in a dense cloud or in pitch-darkness over the ocean. The pilot may be convinced that he is flying straight when, in fact, the plane is in a banked turn, or when coming out of a level turn, he may feel he is diving. Trying to correct the (false) position of the aircraft only makes matters worse. During a rapid deceleration, a pilot sometimes feels the plane is facing downward. To rectify this illusion, the pilot may then pull up the nose of the plane, which often leads the aircraft to fall into a catastrophic spin known, for obvious reasons, as the "graveyard spin." The graveyard spin is what seems to have happened to the Piper plane piloted by John F. Kennedy, Jr. It crashed into the Atlantic Ocean on July 16, 1999, after Kennedy suffered spatial disorientation while flying at night in bad weather en route to Martha's Vineyard.[8]

How can a pilot be convinced that he is flying up when he is actually heading down? Or that he is moving straight ahead when he is, in fact, in a dangerous bank? The human brain's navigational system has evolved to detect our movement on earth, not in the sky. It calculates our position by comparing signals from the inner ear (which has tubes of liquid that shift when we move) to the fixed sensation of gravity that points down to the center of the earth.[9] This system works extremely well when we are on the ground, as it was developed to function in this context (our ancestors did not spend much of their time airborne). However, in a speeding plane in midair, the system gets confused. Our brain interprets irregular signals, such as angular accelerations or centrifugal force, as the normal force of gravity. As a result, it miscalculates our position in relation to the earth.

The liquid in the inner ear does not quite catch up with the fast rate of the plane's directional change, causing false signals to be transmitted to the brain. When our eyes cannot confirm directional change, either, because visual cues are lacking, the change in position can go undetected. The result is that the plane can be flying on its side, while the pilot is utterly convinced it is parallel to the ground; he feels as if he were relaxing on his couch at home.[10]

Now, here is the problem: Throughout life, we have learned to rely on our brain's navigational system to give us the correct position of our body relative to the ground. We seldom suspect it is giving us misinformation, and thus we do not normally second-guess our sense of position. At this very moment, while reading this book, you know for sure that the sky is above you and the ground is beneath. You are probably right. Even in the dead of night, with no visual cues, you can still tell with certainty which way is up.

So the first thing a pilot must learn is that although he may feel 100 percent certain that his plane is going in a specific direction, this may be an illusion. This is not an easy concept to grasp. An illusion is an illusion because we perceive it at face value—as reality. "The most difficult adjustment that you must make as you acquire flying skill is a willingness to believe that, under certain conditions, your senses can be wrong," says one student pilot training guide.[11]

The good news is that there is a solution for a pilot's vertigo; it is the plane's navigational system. This is why, thankfully, most planes do not end up in the ocean, although almost every pilot has had a brush with vertigo at least once in his career. If a pilot is familiar with the plane's navigational system and knows he must rely on it even when it communicates information that contradicts that conveyed by his brain, he will avoid tragedy. The problem in the case of John F. Kennedy, Jr., was that he was not certified in instrument flight rules (IFR), only in visual flight rules (VFR). He was not trained to fly in conditions that did not allow for the use of visual cues—conditions in which one must rely on instruments alone to navigate, such as that dark, stormy night his plane crashed.[12]

Khadr Abdullah, the experienced pilot on the Flash jet, was certified in both IFR and VFR. However, on that fatal day, his brain seemed to trick him into believing he was flying level as he guided the plane into a dangerous right overbank nose-down. How could this happen to an experienced pilot? The U.S. investigative team suggests the following scenario: Shortly after takeoff, the plane was over the Red Sea at night; thus, no visual cues (such as ground lights) were available to indicate ground or sea level. Second, the plane's change in spatial position was so gradual that it could not be picked up accurately by the crew's vestibular systems. In fact, once the angle had greatly increased, the pilot may have perceived that the plane was turning slightly left rather than dangerously right.[13] This scenario is supported by the recordings from the cockpit voice tape. On the tape, the first officer can be heard informing the pilot that the plane is turning right. In a surprised tone, the pilot is then heard responding, "*Right? How right?*" indicating that he has detected a mismatch between the information provided by the first officer and his own perception.[14]

Because of the lack of visual cues and the gradual shift in position, the only way the pilot could have accurately perceived the relative location of the plane to the ground was by constantly monitoring the plane's navigational system. There is evidence, however, that the flight instruments were not being monitored constantly. At the time the plane was entering a right bank, it was allowed to travel at thirty-five knots below the required airspeed and was climbing over the standard pitch. It appears the pilot did not detect these changes because his attention was focused on engaging and disengaging the autopilot.[15] Without monitoring the plane's navigational system, the pilot had only his brain's navigational system to rely on, and that was receiving misinformation from his inner ear and no information from his eyes—resulting in disaster.

Visual Illusions

Most of us have never flown a plane, so we are unfamiliar with the experience of vertigo that can result. Unknowingly, however, we are constant victims of the illusions created by our brain. Take a look at Figure 1, which portrays two squares—A and B. Which one is lighter? You probably see the same as I do: B is lighter. Right?

Figure 1. Checker Shadow Illusion
Edward H. Adelson, 1995.

Wrong. The squares are exactly the same color; I assure you that they are identical. So why do we perceive them as different shades of gray? It is a visual illusion created by our brain. Our visual system believes square B is in shadow, while square A is in light. They are not. The image was created using Photoshop. The squares convey the same amount of light, but our brain corrects for what it assumes to be the position of the squares (in shadow or in light) and concludes that square B must be lighter.[16] The result? Square A looks darker than square B. Our subjective perception of reality differs from objective reality.

Although in this instance our brain has given us faulty information (and in a very convincing manner, too), it has done so for good reason. Our visual system was not built to interpret a cleverly constructed Photoshop image that does not follow physical rules. Like our navigational system, our visual system was developed to interpret the world it would encounter most frequently. To do so, it developed some shortcuts, some assumptions about the world, which it uses to function. These allow our brain to work efficiently in almost all situations. However, it does leave room for errors when those assumptions are not met.

Let's explore another example. Look at Figure 2.

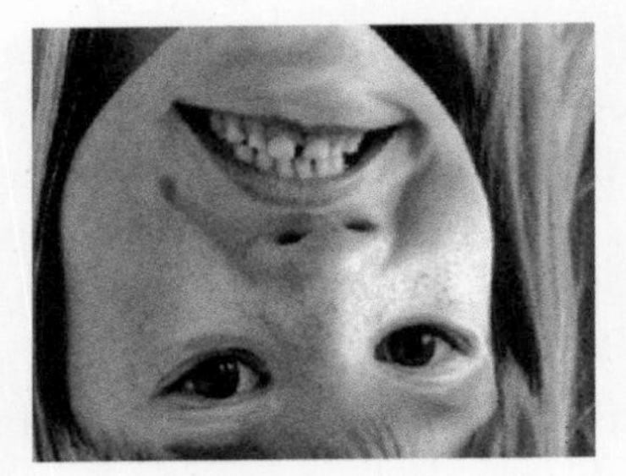

Figure 2. Smiling Girl

Adapted from P. Rotshtein, R. Malach, U. Hadar, M. Graif, and T. Hendler, "Feeling or Features: Different Sensitivity to Emotion in Higher-Order Visual Cortex and Amygdala," *Neuron* 32 (2001): 747–57.

What do you see? An upside-down photo of a girl smiling. Okay, now rotate the book 180 degrees so you can see the photo the right way up. What do you see now? Suddenly, she is not that sweet-looking, is she? The illusion is called the *Thatcher illusion*, as it was first demonstrated in 1980 on a photo of former British prime minister Margaret Thatcher,[17] who, to say the least, is not known for her cheerful expressions.

The illusion is created by inverting a face without inverting the mouth and eyes. Upside down, the face looks relatively normal and the expression perceived is the same as that conveyed by the original photo before it was "Thatcherized" (this is the term for inverting the face without rotating the mouth and eyes). So if the girl was originally smiling, she will be perceived as smiling after being Thatcherized. However, the Thatcherized face looks bizarre when upright, even grotesque. The mismatch between the orientation of the mouth and eyes relative to the rest of the face is easily detected.

This illusion, like many others, gives us clues as to how the brain functions, and the evolutionary constraints that guided its development. We walk around all day encountering upright faces. They are everywhere—on the street, next to us on the bus, or at the office. It is important that we accurately and efficiently recognize that a face is a face rather than, say, a football or a watermelon, because faces really should not be kicked around or split in two. It is also important that you easily distinguish between the face of your significant other and that of your boss or neighbor, as things could get quite awkward if you don't. In fact, just being able to recognize the faces of your partner, boss, and neighbor is not enough. To get along in this world, we need to remember and distinguish thousands of faces. Luckily, most of us do so with ease, thanks to the part of the brain known as the fusiform face area (FFA), which is located in a region of the brain called the fusiform gyrus.[18] The FFA is the part of our visual system that allows us to recognize that a face is a face, and to distinguish between the many faces we encounter on a daily basis. Without a functioning FFA, we may all become *prosopagnosic,* which means we will be face-blind. People who suffer from lesions to their fusiform gyrus have difficulty identifying faces and may even be unable to recognize their own face. (Oliver Sacks famously wrote of such a case in his book *The Man Who Mistook His Wife for a Hat.*)[19]

Imagine living your life without knowing who's who. True, our face recognition is not perfect. We are often approached by people who claim they have met us before but whom we are unable to recall. However, when you fetch your child from school, you usually pick out the right kid, even if he is wearing a new outfit or has just had a haircut. In fact, you do better than that. Not only are you able to detect your child in the mass of faces; you are also able to sense whether your kid had a good or bad day simply by glimpsing the expression on his face.

Humans are very good at perceiving the emotional state of others. We do so unconsciously all the time, using all sorts of clues, such as tone of voice and gait. Mostly, however, we identify the emotional states of others by perceiving their facial expressions. We know a happy expression when we see it on someone's face; we know when someone is sad, afraid, or angry by the exact way his mouth curls and his eyes open wide or become narrow. The clues may be subtle, but we are quite good at detecting someone else's emotional state because we have become experts at identifying facial expressions. We can do so for familiar faces, faces we have not previously encountered, faces from our own culture or a foreign one, because emotional expressions are universal.[20]

The capability to convey and detect emotion is critical to our existence. Take, for example, our ability to differentiate between a fearful face and an angry one. An angry face signals that the person in front of us is upset, possibly at us, and may be a threat to our survival. A fearful face signals that there is a threat somewhere in the environment; however, the person in front of us is not the source of this threat. In this case, we should quickly scan our surroundings to try to detect where the danger is coming from, so it can be avoided.

Accurate recognition of both emotional expressions and identity is vital for social communication. Most of us can recognize thousands of faces; we can easily distinguish Margaret Thatcher from Boy George (apparently, they resemble each other),[21] and a frown from a grin. However, turn faces upside down and we become almost as helpless as a pilot flying in pitch-darkness without navigational instruments.

The brain is used to detect *upright* faces and expressions. It processes the parts of the face (eyes, nose, and mouth) in unison, as this is the most efficient way to do so. In other words, rather than identifying each part separately, the brain processes the face and its expression as a whole.[22] Now, because the brain does not encounter upside-down faces very often, it has not learned to process them as effectively as upright faces. When presented with a rotated face, we seem to process its features separately, rather than in a configural manner.[23]

Let us turn back to the rotated face of the girl in Figure 2. Although her face was rotated, her mouth and eyes were left upright. On their own, the mouth and eyes express emotion in a normal manner. Our brain processes them separately from the rest of the face and identifies the emotional clues conveyed. We thus conclude that the person is smiling. Rotate the Thatcherized face, however, and what is perceived are eyes and mouth in a shape never seen before. The look is deformed, and our emotional reaction to the distortion is disgust and fear.

It's not only humans who are tricked by a Thatcherized face. Monkeys are fooled, too.[24] A group of researchers at Emory University Thatcherized the face of a monkey using the same technique utilized in Figure 2. They then showed a group of monkeys four photos: a photo of a standard monkey face, an inverted photo of a standard monkey face, an inverted Thatcherized monkey face (as in Figure 2), and an upright Thatcherized monkey face (the one humans find bizarre). The monkeys were not very interested in the photos of the standard monkey face—whether the image was inverted or upright, they glanced briefly at the normal face and moved on. What about the Thatcherized face? When the image was inverted (as in Figure 2), the monkeys were no more interested in the Thatcherized face than in the normal face. However, when the image was presented upright, the monkeys spent much longer looking at the Thatcherized face than at any of the other faces. The monkeys' response indicates that they found the upright Thatcherized face as odd as we do, but, like us, they, too, were tricked into perceiving the rotated Thatcherized face as normal. If monkeys are sensitive to the Thatcher illusion, this means that the processes underlying the illusion are evolutionarily old. The brain seems to have developed a specific bias for processing upright faces long ago.

As in most illusions, learning of the illusion and its roots does not erase the illusion. Although we now know that the squares in Figure 1 are the same, we still perceive B as being lighter than A. Our knowledge does not change our perception; the illusion is still there. Similarly, a pilot may acknowledge he is in a state of vertigo, in which the information provided by the instruments does not align with his perception, and still feel that he is climbing up while heading down. The illusion, which feels very real, is dissociated from the knowledge (when available) that the perception is false.

When it comes to visual illusions, it is relatively easy for us to accept that our perception is wrong when it is pointed out to us. We see it with our own eyes. We can rotate the book, or move around the gray squares of Figure 1 in Photoshop, to watch the illusion unfold. However, cognitive illusions, rather than sensory ones, are much harder to accept.

As in any complex system, the brain has built-in defects. These defects are overpowering; we live with them every day without being aware of them. We rarely doubt that our perception is an accurate reflection of the world, when, in fact, our brains can often provide us with a distorted sense of reality. It is when this disparity is pointed out by instruments (as in the case of vertigo), demonstrations (as in the case of visual illusions), or data (as in the case of the optimism bias and other cognitive illusions) that we see an alarmingly different picture from what we expect to see. It is then that we realize that our brains are not quite the final authority on what is around us, or, indeed, within us.

Nevertheless, illusions tell us something about the adaptive nature of the human brain. They convey the success, rather than the failure, of the evolution of our neural systems, but, like vertigo, on occasion they may lead to disaster.

Cognitive Illusions

Glance at the list of personal attributes below. For each of them, take a moment and consider if on that ability you rank within the bottom 25th percentile of the population, the 25th to the 50th percentile, the 50th to the 75th percentile, or above the 75th percentile—in other words, in the top 25 percent of the population.

1. Get along well with others
2. Leadership ability
3. Logical thinking
4. Driving ability

Do the same for these characteristics:

1. Honest
2. Lively
3. Interesting
4. Physically pleasing

The reality is that most people perceive themselves as being superior to the average human being. We think of ourselves as unique. You may have not rated yourself as above average on every single one of the qualities listed. However, I suspect that most of you positioned yourselves in the upper 50th percentile, or even in the upper 25th percentile.

A survey conducted in the mid-1970s revealed that 85 percent of respondents ranked themselves in the top 50th percentile for the ability to get along well with others and 70 percent did so for leadership ability. In fact, for "Get along well with others," a quarter of the individuals thought they should be positioned in the top 1st percentile![25] Another survey showed that 93 percent of the people surveyed believed they were in the top 50th percentile for driving ability.[26]

This, of course, is impossible. *Most* people cannot be better than *most* people. The data depicts a mathematical flaw. Someone has to be in the bottom half of the curve; we cannot all be in the positive tail of the distribution. We can, however, all *believe* that we are at the high end on most positive attributes, and indeed we do. This illusion is known as the *superiority illusion* (or the *superiority bias*). It is as powerful as the illusions produced by spatial disorientation or a Thatcherized face. We are quite confident that we are more interesting, attractive, friendly, and successful than the average person. We may not admit it openly when questioned, but we have a strong sense that this is correct. If truth be told, some of us *are* more creative, honest, and funny than the average person, but about half of us are not. The thing is, we are blind to our own illusions. However, while we do not recognize our own biases, we can often detect biases in others.

This principle applies to spatial disorientation, too. When Khadr Abdullah, the pilot of Flight 604, was guiding his plane into a fatal overbank, he was not alone. Sitting next to him was

Amr Shaafei, his first officer. Shaafei seemed to be aware of the accurate position of the plane. According to the U.S. investigative team's report, "The first officer's verbal communications indicated that he had an accurate awareness of the airplane's flight altitude during the upset sequence."[27] Most likely, when he finally warned Abdullah about the overbank, he was aware of his pilot's vertigo.

For an outsider, someone who is not sharing the illusion that another person is experiencing, the illusion is often evident. Nevertheless, in the case of Flight 604, although Shaafei seemed to be aware of the spatial illusion Abdullah was experiencing, he hesitated in communicating the situation to Abdullah. When he did try to correct the spatial orientation of his superior, it was too late.

This incident demonstrates an important characteristic of many illusions. Contrary to visual illusions, in which we often share the same false perception as those around us, other types of illusions differ slightly according to where we stand. For example, most of us believe we are superior in many ways to other individuals. This means we see *ourselves* as better, not everyone else as better. Therefore, (a) we all have a slightly different view of the world, and (b) we are able to detect cognitive illusions, such as the superiority illusion, in others. Because we can identify these illusions and biases in others but not in ourselves, we conclude that we are less susceptible to bias than most other people. In essence, this means we hold the illusion that we are immune to illusions. This is the irony of cognitive illusions.

Our tendency to perceive ourselves as less susceptible to bias than the rest of the human race was termed the *bias blind spot* by the psychologist Emily Pronin of Princeton University.[28] As an example of this phenomenon, Pronin points to duck hunting.[29]

In 2004, Supreme Court justice Antonin Scalia and Vice President Dick Cheney went on a duck-hunting trip at a private camp in southern Louisiana. I assume both Cheney and Scalia believe they are superior duck hunters—but that is not really the point. The reason the hunting trip is of interest is that Justice Scalia was due to rule on a case in which the vice president was a party. Cheney was appealing the decision of a lower district court, which had ordered him to disclose details about the identity of participants in his energy task force.

The media and the public felt that Scalia should recuse himself from the case, given his close social contact with Cheney only weeks after the Supreme Court had agreed to take on the appeal.[30] The concern was that eating, drinking, socializing, and duck hunting with the vice president might not allow Scalia to be truly objective when later judging Cheney's case. Scalia's response? "I do not think my impartiality could reasonably be questioned," he said, then added that the only thing really wrong with the trip was that the hunting was lousy.[31]

Scalia ended up ruling in favor of Cheney's position, as did the majority of the Supreme Court justices. Although it is possible that Scalia was objective in his ruling, it seems unreasonable to claim that his impartiality could not be questioned. Why does this seem clear to us when it did not to him? Pronin suggests that this is because people tend to judge the extent of *other* people's bias according to their behavior but judge their *own* biases according to their internal feelings, thoughts, and motivations.[32] Scalia went off on a mini-retreat with Cheney, drinking red wine and sharing hunting tips. Soon after, Scalia ruled in Cheney's favor. We evaluate such behavior and conclude that Justice Scalia may have been biased. Antonin Scalia, unlike us, had access to his own thoughts and motivations. He evaluated those and concluded with certainty that he was unbiased when ruling in Cheney's case. Scalia thought he had insight into his inner motives and mental state; he believed he knew which way was up. He was, at least partially, mistaken.

Scalia seems to have experienced an *introspection illusion.* An introspection illusion is the strong sense people have that they can directly access the processes underlying their mental states. Most mental processes, however, are largely unavailable for conscious interpretation. The catch is that people are unaware of their unawareness. Thus, although introspection feels as if we are simply observing our inner intentions, it is largely an *inference* about our inner intentions, rather than a true reflection of them.[33]

One of the best examples of the introspection illusion comes from a study conducted by Petter Johansson, Lars Hall, Sverker Silkstrom, and Andreas Olsson (the last of whom I was fortunate enough to share an office with throughout my Ph.D. studies). The Swedish team set out to examine to what extent intentions are available for accurate introspection.[34] They presented 120 participants with fifteen pairs of female photos. In each trial, the participants had to indicate which of two female photos they found more attractive. They were then given the photo of the chosen female for closer inspection and were asked to explain why they found this woman more attractive than the other. Unbeknownst to the participants, they were tricked by the experimenter during three trials. They were given the photo they had rejected rather than the one they had selected. Amazingly, in about 75 percent of the cases, the participants did not notice the swift switch. This was true even in cases where the two photos were quite different from each other. At the end of the experiment, the naïve participants were asked a "hypothetical" question: "If you were to participate in a study where the photo you had chosen was secretly switched with the one you had rejected, would you notice the change?" Eighty-four percent of the participants (who just moments before had failed to detect any changes) believed they would easily detect the switch.

Even more astonishing was the fact that participants were

more than happy to explain to the experimenter why they found the photo they had actually rejected a few seconds ago more attractive than the one they had, in fact (unbeknownst to them), picked. One participant explained that he had chosen the photo of a smiling girl who was wearing jewelry because "[s]he's radiant. I would rather have approached her at a bar than the other one. I like earrings."[35] In fact, the participant had not selected the smiling girl with earrings! He'd picked the somber one with no jewelry. When fooled into explaining why he preferred the smiling girl, the participant believed he could evaluate the mental processes that had guided his decision. His answer indicates that although he thought he had direct access to his preferences and intentions, he was wrong. He was experiencing an introspection illusion. Instead of truthfully reflecting his inner mental processes, he was inaccurately inferring and constructing his intentions and past mental state.

The researchers dubbed the phenomenon *choice blindness,* and the participants' disbelief that they could be fooled in this way was described as *choice blindness blindness.*[36] The team wanted to make sure that choice blindness was not specific to judging the physical attractiveness of faces. As mentioned earlier, facial processing is special; we process faces in a holistic manner, and it could be that something about face perception is especially susceptible to choice blindness.

Johansson and Hall thus ventured to the nearest supermarket to set up a jam-tasting stand. They stopped unsuspecting customers and asked them to taste two types of jam, black currant and raspberry. One jam was in a blue jar and the other in a red jar. After the customer had a sample of each, he was to indicate which one he preferred. He then received a second sample of what he was told was the jam he had selected and was asked to explain why he preferred it to the alternative.

Unbeknownst to the customers, Johansson and Hall were

playing a little trick once again. The pots of jam in front of the customers were double-ended, with a divider in between, so that each end had a different jam. This made it easy for Johansson and Hall to carry out their magic and give the customer a taste of the jam he had rejected without his noticing. The customers were "jam-blind." They did not notice that the jam they received the second time around was not the one they had chosen. Once again, they confidently explained what had motivated the choices, which they, in fact, had not made.[37] "Less obviously sweet," explained one customer when asked why he preferred the jam he had actually rejected. "Blends well with the plastic spoon," reasoned another.[38]

The experiments by Johansson and Hall and their colleagues show that we can unknowingly create verbal rationalizations for preferences and intentions that we do not actually possess. Does such a disconnect also exist when no deception is involved?

Before making big decisions in life, such as whether to move to a foreign country, which college to attend, or whether to accept one job offer or the other, most of us spend large chunks of time listing the pros and cons of each alternative. We go over the possibilities again and again before finally making a decision. By the time we reach a conclusion, we are ready to explain to anyone willing to listen why Columbia Business School is a better fit for our needs than Wharton. Some of us spend hours debating which movie to watch on a Friday night, and others can make mental lists of the benefits and disadvantages of pepperoni pizza over mushroom and ham before ordering takeout.

Often, we are wasting valuable time. Studies show that thinking too much can lead to suboptimal judgments. In one study, participants were asked to choose, from a few available options, an art poster to take home.[39] One group of participants was asked to list the reasons why they liked or disliked the posters before making up their minds. Another group of participants was allowed to make only snap judgments. When probed a few weeks later, those participants who had made quick judgments expressed greater satisfaction with their selection than participants who had taken the time to consciously evaluate the options.

What's going on? Why does thinking more lead to poorer choices? Conscious assessment of the options caused people to focus on certain aspects of the posters, at the expense of other, more critical aspects. The features that received the greater weight were the ones that were easily verbalized. "The colors of this poster will fit better with my furniture at home," one participant explained. Other elements, such as an emotional response to the poster, were not as easily accessible for introspection and verbalization and thus were more likely to be ignored in the deliberation process. When participants took the posters home, those aspects, the ones they had not really been able to put a finger on, turned out to be most important.

Although it is common to think that deliberation is the finest way to assess which option is best, it may provide us with misinformation. Whether you are selecting apartments or jelly beans, deliberation has been shown to hinder satisfaction.[40] This is because conscious rationalization allows access to only certain data. No matter how hard we try, some mental and emotional processes are likely to remain hidden.

The optimism bias is a cognitive illusion. We are blind to it exactly as we are to visual illusions or the superiority illusion—that is, until hard data is presented. As in choice blindness blindness or the introspection illusion, we do not believe we can be fooled. Yes, maybe our colleagues are unrealistically optimistic, as are citizens of a foreign country, but not us, not Europeans/Middle Easterners/New Yorkers/lawyers/journalists/academics/senior citizens (you get the picture). Many of us think we are fairly realistic about what the future holds, and although our expectations may be relatively rosy, it is because the future, well, is going to be all right.

The optimism bias stands guard. It's in charge of keeping our minds at ease and our bodies healthy. It moves us forward, rather than to the nearest high-rise rooftop. Well, you may say, if that is the case, why would I want to burst my pink bubble? And you have a point. But think back to the Thatcherized face (Figure 2) or the light illusion (Figure 1). You fully acknowledge those illusions, you may even be able to explain how the mind creates them, but you are still fooled. Every time. Every single time.

The same principle holds for the optimism bias. You may acknowledge the bias after reviewing the evidence presented here. At times, this knowledge may even change your actions, just as understanding the mechanisms underlying vertigo allows a pilot to direct his plane safely to its destination. The glass, however, is likely to remain half full.

CHAPTER 2

Are Animals Stuck in Time?

The Evolution of Prospection

It was in 2007 that Jay encountered her fifteen minutes of fame. Her case was first reported in the prestigious scientific journal *Nature* and then in magazines, newspapers, and blogs all around the globe. Psychologists, biologists, neuroscientists, and the general public were all fascinated with Jay's abilities, and the scientists who revealed them became known worldwide.

Jay does not appear to find her skills unusual. As far as she's concerned, her life is far from extraordinary. She lives on the campus of Cambridge University, one of the oldest and most highly acclaimed universities in the world. The university's colleges are situated along the river Cam, which flows peacefully through the city. In one of those beautiful ancient buildings, surrounded by lawns, Jay resides with a number of colleagues. At any given time there are about ten of them sharing a few rooms on one floor. Normally, Jay and her mates get along quite well. There are, however, a few hiccups. In particular, they tend to steal one another's food. Any of you who had housemates in your youth will know that food theft is quite common. You get up in the morning and fill a bowl with crunchy nut cereal, only to discover that you are out of milk. What do you do? Grab the carton your roommate purchased just the other day. While you are getting the fresh milk from the fridge, your attention is captured by a tasty-looking brownie. You know you shouldn't, but the dark cube with nutty pieces is tempting, and after a few seconds of moral debate, you reach for that, too. No one can prove I ate it, you say to yourself. There are five of us living in this house. You quickly stuff the chocolaty square in your mouth, making sure to eliminate incriminating crumbs while licking your lips in satisfaction.

To avoid such unpleasantries, people will often hide their food when living with others, placing the ice-cream carton at the very back of the freezer or the expensive bottle of wine in the bedroom. Jay does the same, and she is quite canny about it, too. If one of her mates sees her storing her breakfast food in a certain location, she will return later, when he has left, and place it in a different site to ensure that the food will not be gone in the morning. At first, this rehiding trick did not occur to Jay. It was only after she herself stole someone else's food that she started engaging in such devious behavior. It takes one to know one.

There is one peculiar aspect to Jay's living arrangements. Jay and her mates do not have assigned rooms. They may go to sleep in any of the floor's quarters. Jay does not mind sleeping in different parts of the residence. This does raise one problem, though. Jay hates being hungry in the morning. Wherever she lays her head, she wants to know that breakfast will be available when she wakes up. Obviously, there is no room service; she lives in a college, not a five-star hotel. So, before going to bed, Jay hides some breakfast food in the room she expects to wake up in the next day. In general, Jay likes her meals to be diverse. If, for example, she knows there are some breakfast grains stored in the bedroom already, she will not bother with more breakfast grains, but will bring some peanuts with her instead.

It is not only for breakfast that Jay enjoys a selection of foods. She prefers an assortment at any meal. If she eats one particular food for a while, she gets bored with it and craves something different. This is quite understandable. None of us likes having the same thing over and over. After a bowl or two of soup, we are unlikely to want another. Even if it is the most delicious soup we ever tasted in our entire life, we are going to want something else next—maybe a salad or a sandwich. The thing is, Jay does not have full control over her daily menu. This is because her meals are catered by the college. Jay has learned early that her need for diversity is not always met by the Cambridge University catering service. Often, exactly the same meal will be offered for both lunch and dinner on the same day. Jay thinks that is appalling. To jazz it up, she will frequently save some of the breakfast food to have for dinner.

There is one more thing I should mention about Jay. Jay can fly. At any given moment, she can spread her wings and take off. But that is not what is fascinating about Jay. It is not why her case was reported in *Nature.*[1] Jay can fly because she is a scrub-jay, a beautiful blue bird, about twelve inches in height and three ounces in weight. The scrub-jay, also known as *Aphelocoma californica,* is a member of the Corvid family and is native to western North America. Jay and her fellow birds were brought to Cambridge University from the University of California at Davis by the experimental psychologist Nicky Clayton.

In the mid-nineties, Clayton, a native Brit, was conducting her postdoctoral research at Davis. One day, while having lunch on Davis's green campus, she noticed the scrub-jays flying around the lunching students and collecting bits of leftover sandwiches. That on its own was not unusual. But the birds did not eat their gatherings there and then. They would hide their treasures around campus, only to return later and rehide the food in yet a second location to eat at a different time.[2]

Most of us would not have given the birds' actions a second

thought. But Clayton was a trained psychologist studying animal behavior. For her, this was a eureka moment. Not only were the birds exhibiting impressive spatial memory, recalling exactly where they had previously cached (i.e., hidden) the food, but by hiding food for later they were demonstrating an ability to plan for a future time when resources might be less abundant. On top of that, their rehiding strategy suggests they were concerned with possible pilfering by other birds. Clayton would prove these hypotheses empirically about a decade later.[3] Her observations seemed to defy an assumption held by many prominent psychologists—the notion that nonhuman animals are mentally "stuck" in time: they cannot imagine being in a different time or place.

I am sitting in my office. It is early fall but already gray and rainy outside. Alas, this is typical London weather. Although I am physically fixed in front of my computer at Queen Square on September 15, 2009, my mind is elsewhere. A moment ago, I was traveling back to Davis, California, 2005, recalling the many lunches I, too, had had on that very lawn, oblivious to the scrub-jays around me. I also remembered the delightful dinner I had had with Nicky Clayton less than a year before, at the oyster bar near London's Borough Market. Clayton does not eat meat or poultry, so we'd opted for seafood and white wine.

While taking a short break from writing, I reserve my flight for an upcoming conference in Chicago. I have been to Chicago only once before, in October 2006. I was on my way to L.A. but missed my connecting flight and had to spend an unplanned night in the Windy City. As I had packed for L.A. weather, I found myself shivering in the cold Chicago night. It was impossible to buy a coat or sweater, as stores had already shut for the day. In a stroke of genius, my friend who was traveling with me popped into a 24/7 convenience store and purchased a fleece blanket. We enjoyed the Chicago nightlife covered in a gray blanket from CVS.

I start making plans for the upcoming trip. I need to prepare my conference presentation, book a hotel. I also must remember to pack more appropriately this time. Walking around town covered in a fleece blanket should probably be avoided. I suspect I will have some time to explore the city. I daydream of what the trip will entail. Although I make detailed plans for the upcoming trip, I will soon find that life has different plans in store. Once I get to Chicago, I will not end up staying at the hotel I originally booked. I will spend my time with people I had not planned to be with, and will learn different lessons from the ones I had expected to acquire.

Mental time travel—going back and forth through time and space in one's mind—may be the most extraordinary of human talents.[4] It is also one that seems necessary for optimism. If we are unable to imagine ourselves in the future, we may not be able to be positive about our prospects, either.

While most of us do not think of mental time travel as a special skill (such as language or arithmetic), we should certainly not take this ability for granted. Our capacity to envision a different time and place is critical for our survival. It allows us to plan ahead, greatly increasing our odds of sticking around this planet. It motivates us to save food and resources for a time when we expect them to be less available. It enables us to endure hard work in the present in anticipation of a future reward, or to search for an appropriate long-term partner. Our voyage is hardly limited to the recent past and the future. It can expand to a time before and after our own existence. This allows us to forecast how our current behavior may influence future generations. If we were not able to picture the world in a hundred years or more, would we be concerned with global warming? Would we attempt to change our actions?

It is easy to see why cognitive time travel was naturally selected for over the course of evolution. But are we the only ones with a capacity for prospection? Do we share this ability with other species? Which ones? Our closest living relatives, the apes? Or maybe evolutionarily distant animals such as fish or birds?

Resolving this issue is made difficult by the fact that animals cannot verbally communicate. In the absence of language, we cannot ask birds, monkeys, and dogs what they anticipate for the future, or what they remember from their past. If Clayton's birds could tell us whether they recall their birth town of Davis, whether they ever think of their jolly days in the sun, get excited by the possibility of flying in the park on the weekend, or imagine themselves getting old, we would have our answers. They cannot. So we are left with carefully observing their behavior in order to infer whether time travel might take place in their minds. Until Clayton conducted her groundbreaking experiments, there was hardly any evidence to suggest that animals can engage in mental time travel. The most prominent hypothesis was the Bischof-Köhler hypothesis, according to which humans are the only species that can imagine the future and mentally reexperience the past.[5]

You may disagree. What about birds migrating to warmer climates? And bears hibernating in winter? Are these examples not evidence of future planning? And what about your dog wagging his tail in anticipation of his next meal when you enter the room? Is that not indicative of both the memory of past events and an expectation of upcoming events?

Not quite. These examples are not what psychologists mean when they refer to mental time travel. Let me explain. Certain animal behaviors, such as storing food or seasonal migration, do not necessarily involve an understanding of future need. These tendencies can simply reflect an evolved genetic predisposition.[6] For example, changes in temperature can automatically prompt migration, without any future planning on the part of the animal. A bird's physiology is wired in such a way that environmental cues trigger a specific action (migration), without anticipation that staying in place will result in cold and unpleasant conditions. Another example is nest building. Birds build nests before ever having eggs to lay in them. Are they anticipating a need to accommodate future eggs? Maybe. But it is likely they are driven by a physiological trigger that is independent of an ability to foresee the future.

Anyone with a dog, a cat, or even a fish knows that the pet can learn. A dog can recognize its owner, be trained to catch a ball or refrain from urinating indoors, and learn that the sound of a can opener means dinner is on its way. Even a fish seems to know that a tap on its bowl will be followed by the delivery of food. There is no doubt that nonhuman animals possess memory. However, the fact that they can associate a stimulus (the sound of a can opener) with an upcoming reward (food) does not mean they can engage in mental time travel. These associations can be acquired in an implicit manner without necessarily involving mental time travel.[7] For example, we know that holding a cup of coffee can be painfully hot, and so we often use a paper sleeve around the cup. Although the knowledge that a hot beverage is hazardous may have been acquired from painful past experiences, we do not need to recall a particular incident in which we had hurt our fingers, or be able to imagine our hand burning, in order to grab a paper sleeve.

Clayton's birds, however, appear to show more than simple associative learning or genetic predisposition. Let's go back to the beginning of this chapter and reevaluate Jay's actions. All the behaviors I described were, in fact, reported by Nicky Clayton and her colleagues. The birds were indeed observed hiding food in places where they expected food to be scarce.[8] They also rehid their food if another bird had previously been watching them, in order to reduce theft,[9] and they would cache a certain food in a location where they expected that particular food to be absent the next day.[10] Would you consider those behaviors indicative of future planning? Do they reflect time travel?

Let's zoom in on a few examples. In one experiment, Clayton had her birds wake up in one of two rooms.[11] In room A, the birds always received breakfast—that was the breakfast room. In room B, they never received breakfast—that was the no-breakfast room. During the day, the birds would hang out in room C and were given plenty of food. They were allowed to eat the food there and then, as well as cache it. What did the birds do? They took some worms and kibble from room C and hid them in the no-breakfast room. Although they were full at the time, they were already anticipating the next day's hunger in the no-breakfast room. The birds' sophisticated behavior reflects specific, elaborate planning. It cannot be explained by genetic disposition and is difficult to account for by stimulus-driven associations. In fact, it is reminiscent of human planning. Just as a bird plans for a time when it will be hungry by moving food from one place to another, humans will go grocery shopping even if they have recently had a large meal, because they anticipate being hungry in a few hours and are aware that the fridge is empty.

That's not all. The birds also seemed to understand the notion of expiration dates. They learned that worms decay faster than pine nuts, and would retrieve worms first if only a short time had passed since they'd hidden them (i.e., if the "expiration date" had yet to elapse), but they would go to the location of the pine nuts if they had calculated that the worms would already have gone bad.[12] This is quite an impressive skill; other animals, such as rats, fail to exhibit an understanding of how time affects food decay.[13] Humans, of course, are well aware of the processes. It demonstrates we can track time, prioritize, and plan for the future. Apparently, scrub-jays can, too.

Obviously, birds are unlikely to have the same level of sophisticated planning as humans do, and are unlikely to imagine the future with the same amount of detail. However, they do not seem to be stuck in the now, either. At least a certain amount of understanding that tomorrow may be different from today is apparent. You might think that if birds can grasp this concept, then surely our closest living relatives, apes, can, too. This does not seem to be the case. There have been several attempts to test whether monkeys plan for the future. Most cases failed to demonstrate such planning in nonhuman primates.[14] When given food, monkeys will eat until they are full and trash the rest. They do this even if they are fed only once a day and will certainly be hungry in a few hours. If they were to save the accessible food, they could avoid being hungry later, but they don't. When they are allowed to choose different amounts of foods (for example, two, four, eight, ten, or twenty dates), they do not always choose the largest amount, but will often pick the amount they can eat there and then. Some researchers have been able to train monkeys to display behaviors that indicate a form of understanding of future time (for example, picking less food now in order to receive more later).[15] But by and large, either monkeys do not have a well-developed sense of future time or scientists have yet to conduct the right experiments.

The Knowledge

So what makes certain birds predisposed to mental time travel? Although birds' basic ability for prospection most likely evolved separately from that of humans, the answer may be found in the brains of London taxi drivers. London cabdrivers are the "top gun" of taxi drivers. To become a licensed driver of the traditional black cab, candidates need to pass an exam in which they demonstrate "the knowledge." "The knowledge" was initiated back in 1865, and it has since become the world's most demanding training course for taxicab drivers. The drivers are required to acquire an intimate acquaintance with 25,000 streets and 320 routes within a six-mile radius of Charing Cross in London's city center. A familiarity with all points of interest along the routes, including theaters, hotels, Tube stations, clubs, parks, and embassies, is mandatory.[16]

If a customer has just been picked up at the National Portrait Gallery (after viewing the latest exhibition, *Twiggy: A Life in Photographs*) and wants to continue to Ronnie Scott's for some jazz (Georgie Fame and the Blue Flames are on), the driver should be able to decide within seconds which is the best route to the passenger's requested destination, taking into account the weather and traffic conditions. He needs to remember which streets lead where, which roads are one-way, which are jammed during rush hour. He cannot be wasting valuable time looking at a map or a GPS or asking for directions on the radio. The knowledge should be stored in his mind, then be retrieved instantly when needed. He must always be a step ahead—taking a right turn while planning for the next left, anticipating the red light before it even appears in sight.

On average, it takes three years of intensive training and about twelve attempts at the final exam to receive a license. Only the best of the best survive—the Mavericks and Icemen of top-gun cabdrivers. This is probably why riding in a London cab is incredibly costly. I suspect New York City taxis are half the price because figuring out how to get from Third Avenue and Fifty-sixth street to Fifth Avenue and Tenth Street is less of a challenge. Barely a quarter of those who embark on the London course are successful; the rest drop out. The ones who pass usually stick with the job for decades, becoming wizards at navigating through the complex streets of London. How do they do it?

Eleanor Maguire, a professor at University College London, scanned the brains of London taxi drivers to find out. While examining the scans, she observed intriguing irregularities. The posterior part of the cabdrivers' hippocampi was larger than average.[17] The hippocampus (there is one on each side of the brain) is a region that is crucial for memory. The posterior bit is particularly important for spatial memory. Taken alone, this finding may indicate that people with larger hippocampi are more likely to become taxi drivers because they are better with navigation—just as taller people are more likely to become basketball players. That was not the whole story, however. Maguire discovered that the drivers' hippocampi grew on the job! They became larger and larger with every year behind the wheel—just as basketball players' calves become larger after years of playing. Drivers with forty years of experience had more gray matter (which contains the neural-cell bodies) in their posterior hippocampi than did newbies who had been on the job for only a couple of years. The cabdrivers' brains had literally made room for the acquired skills and knowledge.

"I never noticed part of my brain growing—it makes you wonder what happened to the rest of it," said David Cohen, a London cabdriver.[18] Indeed, what had happened to the rest? Apparently, while the posterior part of the hippocampus grew, the anterior part was shrinking.[19] The more years on the job, the smaller the anterior part became. The anterior region of the hippocampus is also involved in memory processing, but it is less important for spatial memory. The shrinkage reflects a reorganization of the hippocampus to accommodate the newly obtained skills. The acquisition of these special skills, however, came at a cost. The cabdrivers' extensive knowledge of London was accompanied by impaired memory for other types of information. For example, the cabdrivers performed worse than average when asked to memorize pairs of words (such as *apple* and *toy*). These deficits were not permanent. Once the cabdrivers retired, their brains began changing again.[20] The posterior part of their hippocampi slowly shrank back to its original size, and although their navigation abilities decreased, their scores on other memory tests returned to normal. This is a striking example of how plastic our brains are—ever changing according to our altering needs.

A similar process takes place in the brains of birds. Their hippocampi grow and shrink as a function of how, and when, they are used.[21] Birds that cache food have bigger hippocampi than birds that do not.[22] The volume of the hippocampus is related to the number of locations where a bird stores food and the duration for which it stores this food.[23] Scrub-jays, for example, can store thousands of food items, each at a different location. They can leave their treasures hidden for months and still return to the correct location to retrieve them. (I, on the other hand, often can't remember where I parked my car.) During the fall, when food storing is at its peak, the hippocampi of caching birds enlarge.[24] When the season is over, they shrink back. Just like the hippocampi of the retired taxi drivers, the birds' hippocampi adapt to their needs.

It is not only food-related needs that the birds' hippocampi are sensitive to. They respond to other memory requirements, as well—for example, the necessity of remembering where offspring have been placed.[25] Some birds, such as cowbirds, engage in brood parasitism. This means that they leave their eggs in the nests of others so that the unsuspecting birds will raise their offspring for them, leaving them commitment-free. This is the birds' version of a full-time nanny, only the nanny does not get paid and believes the babies are her own. Before she lays her eggs in a host's nest, the bird conducts a bit of research. She flies around, looking for a suitable nest for her eggs. She then needs to remember the location of the chosen nest so that she can return later to lay her eggs. In one species of cowbirds—the shiny cowbirds—the female bird goes searching for a nest on her own. In these birds, the hippocampi of the females are larger than those of the males, presumably to accommodate additional memory resources. However, in other types of cowbirds, the male and female search for a nest together. In these cases, the size of the hippocampus does not differ with gender.

Similar need-driven hippocampus changes are seen in voles. Voles are small furry creatures, which I discuss at length in chapter 4. They come in two variations: the prairie voles, which are monogamous in nature, and their cousins, the meadow voles, which are polygamists. We all know that having a significant other requires a certain amount of memory power. We need to remember birthdays and anniversaries; likes and dislikes; names of family members, coworkers, and friends. Imagine you have five or even ten spouses. With more partners, there is more information to be stored. There is also a penalty for associating the wrong information with the wrong partner (surely Lucy will not appreciate a birthday present on Nancy's day). In voles, hippocampus size is apparently sensitive to "marital state," and it changes with the number of sex partners. The meadow voles (the ones that sleep around) have larger hippocampi than their monogamous cousins. What's more, the size of their hippocampi correlates with their home range—voles that have multiple partners that are geographically scattered have larger hippocampi than voles with partners that are all located near one another.[26] The greater the distance to be traveled between female partners, the greater the size of the hippocampus. Presumably, a larger hippocampus supports better spatial memory, allowing the voles to navigate successfully from one lover to the next. Does it also support a grasp of future time that helps to accommodate multiple encounters?

Our Ability to Travel Through Time

The term *mental time travel* was first coined by the Canadian psychologist Endel Tulving to refer to our capacity for revisiting the past *and* imagining the future. Tulving claimed that these two abilities are related: They rely on the same cognitive and neural mechanisms.[27] In 1985, he reported the case of K.C., an amnesiac patient who not only found it impossible to remember his past but also was unable to say what he expected to do in a year, a week, or even the next day. When asked about his past or future, K.C. claimed his mind was blank. K.C. had suffered damage to his frontal and temporal lobes, including a lesion in his hippocampus. Two decades later, Eleanor Maguire (the scientist who conducted the experiments on London cabdrivers) examined amnesiac patients with brain damage confined to their hippocampi. She found that those patients, just like K.C., were not able to construct detailed images of future scenarios.[28] Without working hippocampi, the patients appeared to be stuck in time—unable to revisit the past or mentally explore the future.

Around the same time, a series of brain-imaging studies conducted at Harvard University by the psychologists Donna Addis and Daniel Schacter showed that the hippocampus is engaged when we recollect our past and when we imagine our future.[29] They suggested that the hippocampus evolved not to form and retrieve memories, as previously thought, but to simulate the future.

Member Shuttle ID card required to purchase this pass.

M

Please keep this pass when the last ride is used—it is your receipt.

This pass is non-refundable and does not expire.

To infer what may happen in the future, we need to access pieces of stored information. Hippocampal function has a role in all these processes—encoding the episodes of our lives, storing that information, and retrieving it, as well as imagining the future. It is critical in binding pieces of information together to create a mental image of both the past and the future.

It is not surprising, then, that species, such as scrub-jays, that have evolved to have extraordinary memory also exhibit an ability for prospection. This raises an interesting question: If certain birds show basic signs of mental time travel, do they, too, wear rose-tinted glasses? Once again, answering this question requires a sophisticated approach. We cannot ask birds directly if they anticipate a long, healthy life. We need other methods. Melissa Bateson and her team at Newcastle University came up with one.[30] They trained birds to press a blue lever whenever they heard a short two-second (2-s) tone, at which point they would receive an immediate food reward. The birds were happy to receive a treat and quickly associated the short sound with a positive result. They also trained the birds to press a red lever when they heard a ten-second (10-s) tone, in which case food would be delivered, but only after a delay. The birds were not happy waiting around for their meal (just as we dislike arriving at a restaurant and being told that our table will not be ready for half an hour). The long 10-s tone was therefore associated with a negative outcome.

The birds quickly learned that a 2-s sound meant they should press the blue lever to receive an immediate reward; a 10-s tone meant they had to press the red lever to receive a delayed reward. The birds had to get it right. If they made a wrong response, they would not be given any food at all. Now, the question was, what would happen when the birds heard an ambiguous sound, one that was between two and ten seconds in length? Would they expect a good outcome when they heard a six-second (6-s) sound and press the blue lever? Or would they be pessimistic, expecting an annoying wait, and press the red lever?

The birds showed a positive bias. When they heard an ambiguous sound (such as a 6-s sound or even an 8-s sound), they were more likely to classify it as indicating a good result. Although they had no real reason to do so, they would press the blue lever in anticipation of an immediate reward. There was one catch, however. Only "happy" birds that were living in "deluxe" cages exhibited optimism. These were privileged birds that were housed in large, clean cages. They had branches to play with, access to comforting baths, and a constant supply of water. Birds that were less privileged, that lived in smaller cages, without playthings, with unpredictable access to water and baths, were more realistic. They did not exhibit an optimism bias, and, in general, they classified the tones more accurately. Just like humans suffering from mild depression, birds living in difficult conditions exhibited *depressive realism* (more on the link between depression and optimism in chapter 6): They had an accurate view of the world they were living in that was unaffected by positive illusions.

The vital difference in the level of sophistication of future thinking between humans and birds (and all other animals), however, lies in our frontal lobes. Relatively large frontal lobes are what distinguish humans from their less developed ancestors and all other animals. The rapid development of human frontal lobes allowed for the ability to make tools, find novel solutions to old problems, plan steps that would make goals more achievable, see far into the future, and, most important, possess self-awareness.

While the capacity for both awareness and prospection has clear survival advantages, conscious foresight also came at an enormous price—an understanding that somewhere in the future, death awaits us. This knowledge—that old age, sickness, decline of mental power, and oblivion are around the corner—is less than optimistic. It causes a great amount of anguish and fear. Ajit Varkil, a biologist at the University of California at San Diego, argues that the awareness of mortality on its own would have led evolution to a dead end.[31] The despair would have interfered with daily function, bringing the activities and cognitive functions needed for survival to a stop. Humans possess this awareness, and yet we survive. How?

The only way conscious mental time travel could have been selected for over the course of evolution is if it had emerged at the same time as false beliefs.[32] In other words, an ability to imagine the future had to develop side by side with positive biases. The knowledge of death had to emerge at the same time as its irrational denial. A brain that could consciously voyage through time would be an evolutionary barrier unless it had an optimism bias. It is this coupling—conscious prospection and optimism—that underlies the extraordinary achievements of the human species, from culture and art to medicine and technology. One could not have persisted without the other. Optimism does not exist without at least an elementary ability to consider the future, as optimism is by definition a positive belief about what is yet to come, and without optimism, prospection would be devastating.

CHAPTER 3

Is Optimism a Self-Fulfilling Prophecy?

How the Mind Transforms Predictions into Reality

Champagne corks were popping in the Los Angeles Lakers' locker room. It was June 1987 and the Lakers had just won the NBA championship after defeating the Boston Celtics 4 to 2. For the Celtics, this would be their last appearance in an NBA final until 2008. For the Lakers, other things were in store. The Lakers' 1987 team was one of the best basketball teams of all time. The team included renowned players Magic Johnson, James Worthy, and Kareem Abdul-Jabbar. It was their head coach, Pat Riley, however, who was about to make history that evening.

In the midst of the postvictory celebration, Riley was approached by a reporter. The journalist wanted to know if Riley believed the Lakers could be the first team in almost twenty years to win the NBA championship twice in a row. The Boston Celtics had been the last to do so, back in 1969, but no team had managed a repeat since. Would the Lakers be able to achieve it by winning the championship again in one year's time?

"Can you repeat?" the reporter asked Riley. Without blinking an eye, Riley replied, "I guarantee it." The reporter was blown away. He had to make sure he had heard right. "Guarantee?" he said. "That's right," said Riley.[1] With those three words—"I guarantee it"—Riley promised the journalists, the players, and millions of fans a second championship.

Riley's guarantee was not a champagne-induced slip of the tongue. During the team's victory parade in downtown L.A., immediately after making his initial promise, Riley once again guaranteed the crowds a repeat championship, and he continued to promise a repeat again and again throughout the summer and into the 1987–1988 season. "Of all the psychological things that Pat's come up with, this is probably the best," said Magic Johnson in an interview in 1987.[2]

The Lakers did well that season, and a year after Riley's guarantee, they found themselves yet again at the NBA finals. This time, they were up against the Detroit Pistons' "bad boys," who were hungry for their first title. The battle was tight. The Detroit Pistons moved ahead, winning the first game. Although the Lakers took the next two games, the Pistons took games four and five, and so by game six the score was 3 to 2 in the Detroit Pistons' favor.

With fifty-two seconds remaining in game six, the Pistons were leading 102 to 101. The Lakers turned up the heat, forcing the Pistons' star player, Isiah Thomas, into a hopeless shot, after which Kareem Abdul-Jabbar got the ball. While Abdul-Jabbar was shooting, the Pistons' Bill Laimbeer was whistled for a controversial foul. Abdul-Jabbar made the two free throws successfully, winning the game and bringing the final back to an overall tie, with three wins each for the Pistons and the Lakers.

Game seven would determine whether Riley's guarantee was to be fulfilled. The Pistons were ahead at halftime, but the Lakers turned their shortfall into a lead during the second half. With six seconds to go, they were ahead by a very small margin, 106 to 105. During those six seconds, they managed to score once more and won the last game 108 to 105, making good on Pat Riley's promise.

Seconds after delivering the guaranteed victory, Riley was once again in front of the cameras. The fans were roaring. "Can you do it again?" the reporters asked. "Will there be a three-peat?" Riley opened his mouth to answer the question, but before the words could leave his mouth, the 225-pound Abdul-Jabbar made one of his famous jumps. This time, however, he was not aiming at the basket. He was aiming at Riley's lips, which he managed to cover with his big hands just in time to prevent Riley from making any more guarantees. Abdul-Jabbar would explain later that the pressure to make good on Riley's promise was too much for him to take for yet another year.

Riley never guaranteed a "threepeat." The following season's final was a rematch of the previous year's championship round between the Los Angeles Lakers and the Detroit Pistons. This time, however, the Pistons won the series in a four-game sweep, following which Kareem Abdul-Jabbar announced his retirement at age forty-two. Would the Lakers have won a third consecutive championship if Riley had promised it? We will never know.

Prophecy or Cause?

There were many factors that led the Lakers to win the final against the Pistons in 1988 and lose it the following year. However, it is tempting to speculate that Riley's promise during the 1987–1988 season, and the lack thereof during the 1988–1989 season, had a crucial role in the final turn of events.

Riley's guarantee of a repeat is a classic example of a self-fulfilling prophecy—a prediction that causes itself to be true. There is no doubt that Pat Riley had good reason to think his team would win the championship the following year when the reporter asked him about it after the 1987 game. His team had just won the final, they were declared the best, and thus they were a likely candidate for the next year's championship. However, his statement, which conveyed unshaken optimism, triggered a process that made that guarantee much more likely to become true. "Guaranteeing a championship was the best thing Pat ever did. It set the stage in our mind. Work harder, be better. That's the only way we could repeat. We came into camp with the idea we were going to win it again, and that's the idea we have now," said the Lakers' Byron Scott back in 1988.[3]

Believing that a goal is not only attainable but very likely leads people to act vigorously in order to achieve the desired outcome. In Riley's case, he did more than just predict a repeat; he guaranteed it. By promising a second championship, he piled extra pressure on himself and his players. The Lakers could not disappoint their fans, who were expecting a win; they had to prove their coach was right. So Magic, Kareem, and their teammates had to train harder than they ever had and be better than they ever had been in order to make Pat Riley's prediction come true.

The idea behind the self-fulfilling prophecy is that it is not a *forecast* of a future event, but a *cause* of the event. Don't get me wrong: Predicting your team will win the championship does not necessarily make it so. It is not magic. Not all athletes who envision their success will actually go home with a championship cup or a gold medal. Many factors will determine the outcome, and the opposing team may be just as confident. However, a prediction has an influence on the event it predicts because people's behavior is determined by their subjective perception of reality, rather than by objective reality. Therefore, believing in a positive outcome will enhance the probability that the desired outcome will be realized.

The term *self-fulfilling prophecy* was coined by the sociologist Robert Merton in 1948. According to Merton, "The self-fulfilling prophecy is, in the beginning, a *false* definition of the situation evoking a new behavior which makes the originally false conception come *true.* The specious validity of the self-fulfilling prophecy perpetuates a reign of error. For the prophet will cite the actual course of events as proof that he was right from the very beginning."[4]

Let's use Merton's terms to dissect Riley's self-fulfilling prophecy of the Lakers' repeat. Riley's statement that the Lakers were certain to be the champions the following year was false at the time it was made, as no event is predetermined. Since the future is always uncertain, no one can know for sure what it holds. Thus, Riley's assertion was "a *false* definition of the situation." However, by "evoking a new behavior"—rigorous training and lack of compromise—the claim resulted in a repeat championship, making "the originally false conception come *true.*" After delivering on his promise, Riley may have believed that he was correct all along, since "the prophet will cite the actual course of events as proof that he was right from the very beginning." It is, however, the prophecy itself that made the outcome highly probable.

The self-fulfilling prophecy is an extremely powerful phenomenon. Expectations have been shown to affect everything from education, racial biases, and financial markets to health and well-being, and even premature death. A self-fulfilling prophecy also turned a German stallion into a math wizard.

Communicating Expectations

On September 4, 1904, an article entitled "Berlin's Wonderful Horse: He Can Do Almost Everything but Talk" was published in *The New York Times.*[5] The wonderful horse was a German stallion named Hans, later known as "Clever Hans." Hans's owner, Wilhelm von Osten, was a schoolteacher, who upon retirement decided to teach the horse math and German. Von Osten believed that by using the same techniques he had mastered in the classroom, he would be able to educate anyone, even a horse. For four years, von Osten stood in front of a blackboard and taught Hans to read, spell, do arithmetic, name the date, tell time, and much more. Hans would use his front right foot to answer questions. In response to complicated math problems, for example, Hans would respond by tapping his foot the number of times that corresponded to the correct answer. He created words by using a similar system: Von Osten had the alphabet written out on the blackboard, and Hans used a specific tapping sequence to indicate each letter. Hans could recognize people and spell out their names, using the correct spelling of the complicated German language.

Experts and laymen alike were impressed. *The New York Times* made clear that "the facts in this article are not drawn from imagination, but are based on true observations that can be verified by . . . scientific and military authorities."[6] Could a horse really add, multiply, tell the exact hour, and spell? Or was he simply taught some clever tricks, such as those used to train performance animals in the circus? People everywhere were intrigued and wanted an answer. Did Hans really possess incredible mental powers?

On October 2, 1904, almost a month after publishing the original story, *The New York Times* came out with a follow-up piece: "Clever Hans Again: Expert Commission Decides That the Horse Actually Reasons."[7] The commission, comprised of doctors, zoologists, physiologists, and circus trainers, was unanimous in its opinion: Hans had not been taught fancy circus tricks; he'd been trained by the same methods used to educate schoolchildren. He seemed to have genuine mental abilities.

Hans could give correct answers even if his master was not the one asking the questions, and he could even answer questions about subjects he had never been taught by his trainer. And so the conclusion was that if animals were to be treated as humans, they would think like humans, too.

The commission passed its findings to the psychologist Oskar Pfungst. Pfungst, however, was not convinced. He decided to examine the genius horse himself. Pfungst's careful investigation revealed that the horse could answer almost any question correctly as long as (a) the questioner knew the answer himself and (b) the horse could see the person asking the question. When these two conditions were met, Hans got 89 percent of the questions right. However, when one or both of the conditions failed to be fulfilled, Hans's ability to answer questions dropped to a mere 16 percent. What was going on?

What Pfungst concluded was that Hans was actually responding to unconscious bodily cues from the questioner. The horse would be asked a question and then start tapping away. When the number of taps approached the correct answer, the questioner involuntary tensed his body posture and facial expression in expectation. When Hans delivered the last expected tap, the questioner would change his posture and facial expression again, releasing tension. That was Hans's cue to stop tapping, and so he did. In other words, Hans did not know the answers to the questions. He was not performing complicated math tasks; he did not know if it was January or December, Monday or Thursday. In four years of tedious training, what Hans learned was how to respond to his master's bodily cues in a way that made his master happy. For a stallion, that's not an accomplishment to be dismissed.

Because von Osten believed he could teach Hans language and math and thought he could teach his horse to give correct answers to verbal questions, his expectations had an impact on the horse. True, Hans was not performing the tasks the way his master thought he was. The horse was not retrieving the information von Osten had taught him. Rather, Hans had been (unknowingly) conditioned to stop tapping in response to certain bodily movements of the person in front of him. Yet the end result was the same: The stallion was providing the right answers—the answers von Osten expected him to provide.

When it comes to humans, the influence of socially communicated expectations runs deeper. In the late 1960s, a collaboration between the Harvard psychologist Robert Rosenthal and the principal of an elementary school in San Francisco, Lenore Jacobson, led to a fascinating demonstration of the self-fulfilling prophecy.[8] The two wanted to examine how teachers' expectations affect their students' performance. Could students' achievements be modulated by their teachers' preconceptions, even when those were based on nothing at all?

Rosenthal and Jacobson randomly selected students in Jacobson's school and told their teachers that those children were found to be at a point of significant intellectual growth. This information was false—there was no data indicating that the abilities of those students were different from those of any of the other students.

Nevertheless, at the end of the year, the sham predictions had become reality. The students who had been singled out (at random) by Rosenthal and Jacobson as intellectually superior scored higher on the end-of-year IQ tests than the students who had scored similarly to them at the beginning of the year. Their improvement throughout the year was greater than what would have normally been expected. Just like Magic Johnson, Kareem Abdul-Jabbar, and the stallion Clever Hans, the students delivered what was expected of them.

The conclusion was clear: Humans are hugely affected by the expectations placed upon them. Your employee will be more productive if you expect him to be; your spouse will be more loving if you anticipate her to be; your child will be more likely to do well in school and sports if you believe he is talented, and less likely to do well if you hold negative expectations regarding his ability. Even teenagers' alcohol consumption has been shown to be influenced by parental expectations.[9]

What exactly did Jacobson's teachers do that resulted in the observed enhancement of the academic ability of the students? Rosenthal identified a number of behaviors expressed by the teachers that could have influenced the children's performance: The teachers spent more time with the "talented" students than with the other students, provided them with more detailed feedback, and were more likely to encourage them to respond in class. Overall, the teachers treated the "special" kids differently, and as a result, those kids indeed became special. Rosenthal and Jacobson referred to their findings as the *Pygmalion effect,* after George Bernard Shaw's play. Shaw's *Pygmalion* is a classic makeover tale—the story of a professor who transforms a working-class girl into an upper-class lady.

In Rosenthal and Jacobson's Pygmalion study, preexisting notions were assigned to the children at random. In real life, however, educators, as well as the rest of us, hold relatively stable preconceptions that are, in general, not based on real evidence. Teachers have been shown to form predictions regarding new students' achievements based on race, gender, ethnicity, socioeconomic level, and even physical attractiveness.[10] This can be dangerous. Expectations, as we have just learned, are likely to influence a child's performance, ultimately altering his future. In fact, the Pygmalion effect is thought to be a significant factor in producing and maintaining gender and racial gaps in IQ testing, GPAs, and college success.

The Power of Stereotypes

Stereotypes are just another example of a self-fulfilling prophecy. They are especially powerful in shaping an individual's reality, particularly when the same expectations are held by a large number of people. People conform to their group stereotype because society interacts with an individual in a way that is consistent with the stereotype-based prediction. Take Tom and Rob, for example. Both are students in an elementary school in Washington, D.C. They are about the same height and the same weight, both are average students, and both are well liked by their peers and teachers. Tom is black and Rob is white. Initially, Tom and Rob possessed similar physical abilities. Rob could run as fast as Tom, could jump as far, and was just as prone to shoot the ball into the basket. Everyone, however, expected Tom to be a better basketball player than Rob simply because of the color of his skin. For that reason, Tom was more likely than Rob to be selected to play on the basketball team by his peers. His coach took special notice of him, making sure to correct his game. Tom's parents encouraged him to train extra hours on the court after school. As a result, Tom actually became a better player than Rob. Although the belief that Tom was a better player than Rob was false to begin with, the stereotype became self-fulfilling. Consequently, Tom is another example of the stereotype that blacks are better at basketball, strengthening a preconception that essentially feeds on itself.

Stereotypes nourish themselves not only because they affect the way people act toward the individual being stereotyped but also because individuals have a strong tendency to rapidly conform to what is anticipated of them. The most astonishing example of such rapid fulfillment of expectations is the famous case of

Jane Elliott, a third-grade teacher in Iowa. Her class, like most of those in the state of Iowa, was composed solely of white students. It was April 5, 1968, the day after the assassination of Martin Luther King, Jr., and Elliott wanted to demonstrate to her students how it feels to be racially discriminated against. She came up with a "game." As part of this game, the kids were to be separated into two groups according to the color of their eyes. Elliott announced that blue-eyed children were inferior to brown-eyed children. She said they were less intelligent, were slow learners, and were therefore to be treated differently. Blue-eyed children could break for recess only after all the brown-eyed kids had left the class. Furthermore, they were not allowed to communicate with the superior brown-eyed kids.

Immediately following Elliott's announcement, the kids' behavior changed. Brown-eyed children became more confident, while blue-eyed kids became frightened. Most astonishing was the sudden change in the kids' reading and writing abilities. Elliott observed abrupt improvements in the abilities of brown-eyed kids and a decline in the abilities of the blue-eyed children. The next day, Elliott reversed the roles. Now blue-eyed kids were proclaimed superior to brown-eyed kids. Within minutes, the children adapted to their new roles. Blue-eyed kids were now acting dominant, while the brown-eyed kids were acting scared and frustrated. A spelling test that Elliott conducted on both days revealed that the children did better on the day that their group was declared superior.

Jane Elliott's students were responding to expectations that were explicitly expressed by an authority figure. When told they were stupid, they immediately acted stupid. When told they were smart, they acted smart. One might argue that this is because children are especially suggestible. Surely adults with a well-developed self-image would not be as gullible—or would they?

Not only do adults rapidly conform to expectations in a manner similar to that of Elliott's students; they do so even when they are not consciously aware of what is expected of them. My colleague Dr. Sara Bengtsson, a cognitive neuroscientist, wanted to see if she could influence performance on a cognitive task by priming college students with words such as *smart, intelligent,* and *clever,* or with words such as *dumb, stupid,* and *daft.* After presenting the participants with a word such as *clever* or a word such as *stupid,* she asked them to complete a variety of cognitive tests. She found that students performed better after being primed with the word *clever* than when given the word *stupid.*[11] Bengtsson could influence people's expectations of their own performance of a task simply by presenting them with information that was irrelevant to their true ability. Nevertheless, this irrelevant information unconsciously modulated people's expectations of themselves, and thus changed their performance.

In a similar manner, when individuals are reminded of their membership in a group (such as gender or race), the stereotype associated with that group is more likely to influence their behavior. For example, females have been shown to score lower on math tests when reminded of their gender before the test.[12] Reminding females of their gender unconsciously activates the stereotype that women do not do well in math. This priming lowers females' expectations of their ability to do well on the exam, causing them to accept the stereotype and do worse. In another example, African Americans were found to do significantly worse on IQ-type tests, compared to Caucasians, when race was emphasized, but they did as well as Caucasians when no stereotype threat was present.[13]

What Sara Bengtsson's study shows, as well as Jane Elliott's field experiment, is that the influence of stereotypes is surprisingly fluid. New expectations can rapidly take over old ones, quickly substituting one behavior for another. This fluidity is encouraging. It means that with guided intervention, we may be able to reverse the negative effects of stereotypes on an individual's performance.

Learning from Failed Predictions

How do expectations influence the workings of the human mind? Bengtsson's study provided some intriguing answers. She conducted her *clever/stupid* priming study on volunteers while their brains were scanned in a functional magnetic resonance imaging (fMRI) scanner. An fMRI machine not only shows an anatomical image of the brain; it provides data that can tell us how the brain functions. When neurons in a specific part of the brain are active, their consumption of oxygen is increased. In response, blood flow will be enhanced to that region, supplying hemoglobin (the "oxygen-storage molecule" that releases and absorbs oxygen). This leads to local changes in the concentration of deoxyhemoglobin and oxyhemoglobin, which alters the MRI signal that is recorded by the scanner.[14] So if a subject lies in an fMRI scanner and listens to Rachmaninoff's Piano Concerto no. 2, the Black Eyed Peas, or Al Green, the scanner will record changes in the blood-oxygen-level-dependent (BOLD) signal from the auditory cortex, indicating that activity in that part of the brain is enhanced when listening to music.

What Bengtsson's data showed was that participants' brains responded differently when they learned they had made a mistake, according to whether they previously had been primed with the word *clever* or with the word *stupid.* When participants were primed with the word *clever* and then made an error, enhanced activity was found in the medial prefrontal cortex. This heightened activity was not seen after people gave a correct answer; neither was it observed when they previously had been primed with the word *stupid.*[15]

The frontal cortex is a large area of the brain and includes anatomically and functionally distinct subregions. It is the most recently developed part of the brain, and it is not found in animals at the lower end of the evolutionary scale. Although many nonhuman animals have frontal lobes, theirs are considerably less developed than the frontal lobes of humans. The frontal cortex has enlarged disproportionately in human evolution relative to the rest of the brain. Its physical development is the main reason why we have a relatively larger brain than most other animals.[16]

The frontal lobes are critical for functions that are uniquely human, such as language and *theory of mind.* Theory of mind is our ability to think about what other people are thinking. When you are thinking about whether your supervisor knows that you missed your deadline because you went out last night rather than stayed in to finish your assignment, you are engaging in theory of mind. Theory of mind includes contemplating what other people know, assessing other peoples' motivations and feelings, and considering what others expect from you.[17]

Theory of mind is just one process that requires frontal-lobe function. The frontal lobes are involved in many other high-level mental processes, including *executive functions.* Executive functions are those that enable us to identify future goals and recognize the actions that will move us toward achieving those goals. The ability to predict which behaviors will lead to which outcomes ("Going out tonight will cause me to miss my deadline"), differentiate desired outcomes from unwanted outcomes ("Submitting the project on time is good; getting fired is bad"), and promote actions that lead to the sought-after results ("Stay home tonight to work on the assignment") are all functions that require adequate working of frontal regions.[18]* It is often the case that conflicting inputs will be transmitted, and the frontal lobes are required to differentiate between these conflicting desires, making sure to inhibit actions that are anticipated to lead to less desired results or socially unacceptable outcomes.

We are confronted with conflicting needs and information all the time. After a long day at work, we may want to go home and immerse ourselves in junk TV and eat a bowl of salt-and-vinegar potato chips. At the same time, an inner voice tells us we should go to the gym.[19]† Finding a solution to this conflict involves recognizing the future consequences resulting from the different actions, and directing thoughts and actions in accordance with these internal goals. A well-functioning frontal lobe will inhibit the action that is associated with the less desirable goal and promote the action that is associated with the more desirable one.

In Bengtsson's study, after being primed with the word *clever,* participants expected to do well. However, when they provided an incorrect answer, the outcome was at odds with their expectation. The result (an error) conflicted with the anticipated outcome (doing well) and generated a mismatch signal in the frontal cortex. When the brain doesn't get what it expects, it frantically tries to figure out what went awry. The signal in the frontal cortex may have been modulating attention, signaling "Take notice—something here is wrong."[20] The importance of this signal is that it can facilitate learning. As learning from errors is critical for directing our behavior toward optimal functioning,

*The frontal lobes do not carry out all of these complicated computations alone. They receive important input regarding the value of actions and outcomes from many other regions—most notably, subcortical areas that are involved in processing motivation and emotion, such as the striatum and the amygdala.

†The two desires are at odds. This conflict is thought to be signaled in an area of the frontal lobe called the anterior cingulate cortex.

enhanced attention to errors would lead to better performance on the next trial.

However, when the participants were primed with the word *stupid,* there was no heightened activity in the frontal cortex following an incorrect answer. The participants *expected* to do poorly, and thus they did not show signs of surprise or conflict. Lacking a "Take notice—wrong answer" signal in the brain, the participants failed to learn from their mistakes and were less likely to improve over time. They accepted their errors because errors were expected, and they did not try to regulate their behavior to achieve better performance.

In general, our frontal lobes mediate planning and action to achieve the goals we have identified for ourselves. These may be short-term goals, such as completing the *Times* crossword puzzle or making a gourmet dinner for friends. They may be medium-term goals, such as running a marathon in four hours or learning to play the guitar. They can also be long-term goals, such as being successful at a job, being a good parent, or being happy. Our progress toward these ideals is monitored by matching what we do with what we expect. When we stray from the path we have set for ourselves, when behavior does not quite align with expectations, thoughts and actions are quickly generated to get us back on track. If we expect to get a promotion at work but find ourselves stuck in the same position for years, with no change in sight, we may pause to mull over exactly what went wrong. We may then reassess our behavior, identifying new actions that can lead to our wished-for outcome. Maybe we will be motivated to put in extra hours, or ask for more responsibility. Eventually, these actions can lead to the sought-after promotion we anticipated for ourselves.

However, if we do not expect to get promoted, we will continue with our normal routine. We will not be surprised by a lack of promotion and will not notice that we are not getting ahead. The brain will receive what it anticipated, and no error-related signal will be generated in the frontal cortex to modulate behavior. We will not attempt change, nor will we achieve it.

You may say, "Well, positive expectations may lead to better outcomes, but what happens when they don't?" Surely we can't always get what we want, and, with all due respect to Mick Jagger, we often don't even get what we need. What then? Will great expectations just lead to disappointment? Is it not better to hold lower expectations, thereby protecting ourselves from frustration?

The notion that holding low expectations will protect us from disappointment is known as *defensive pessimism.* Low expectations, however, do not diminish the pain of failure. Not only do negative expectations lead to worse results; they also fail to protect us from negative emotions when unwanted outcomes occur. For instance, students who had low expectations of their performance on an undergraduate psychology exam felt just as bad when those expectations came true as students who expected to do well.[21]

In fact, negative expectations can—literally—kill us. Take Peter and James. Peter is a forty-year-old investment banker, and James is a forty-two-year-old corporate lawyer. On a bright Sunday morning, they both find themselves in the ER following a heart attack. The severity of each man's condition is assessed similarly at first, and they are given the same prognosis. Peter usually sees the glass as half full, and James tends to see it as half empty. The reaction each man has regarding his illness is consistent with his general view of life. Peter believes he is strong and, with a bit of effort, will recover, getting back to his daily routine in no time. James, on the other hand, is convinced his time has come. At the very best, he believes, he will live for a couple more years, and even those will not be of the quality he has been accustomed to.

At the time, there is no evidence to support either prediction: There is no objective reason to think Peter will enjoy a swift recovery and that James's life will end in a quick journey to the morgue. However, their predictions are very likely to influence the outcome of their conditions by altering their behavior, and thus be self-fulfilling. Peter is more likely to take actions that will result in his expected recovery (avoiding fatty foods and salt, avoiding stress-inducing situations, and engaging in moderate exercise). James, expecting his life to be over soon, will be less motivated to do so. He will be more likely to relapse, thus potentially making good on his own prediction of an early death.

In effect, people who react to illness with passive acceptance of their own impending death, such as James, die prematurely.[22] Although James and Peter are the creations of my imagination, they could have easily been two participants in a study conducted in 1996. This study examined a group of patients who had experienced heart attacks and were following a rehabilitation program. The researchers found that, just like Peter, optimists exercised more and were more likely to reduce their body-fat levels, thereby reducing their overall coronary risk. They were also more likely to take vitamins and eat low-fat diets.[23] The result: Optimists lived longer.[24]

Pessimists, on the other hand, die younger. A study tracking one thousand healthy people over fifty years found that pessimists were more likely to encounter an early death than optimists. What killed these poor naysayers? Apparently, pessimists were more likely to perish prematurely as a result of accidental and violent events—car crashes, drowning, work accidents, and homicide were more often in their cards. Why would a bleak outlook result in such tragic deaths? It seems that a pessimistic outlook promoted risk-taking behavior because the pessimists believed they did not have much to lose.[25]

Optimists envision a glorious future and are reluctant to disappear into oblivion. The secret to the positive relationship between optimism and health is that optimists are selective risk takers. They take risks if the potential health problem is trivial and/or unlikely to affect them.[26] They will not avoid talking on cell phones for fear of a brain tumor, as the relationship between the two is not well established. On the other hand, they are less likely to smoke, as smoking is a well-documented cause of lung cancer.[27] In other words, optimists save their mental and physical resources for significant threats.

By definition, optimists are people who hold positive expectations of the future: They expect to do well in life, have good relationships, and be productive, healthy, and happy. Because optimists expect to do better and be healthier, they have fewer subjective reasons for worry and despair. The result? They are less anxious and adjust better to stress factors such as abortion or childbirth, cancer or AIDS, and even medical or law school.[28] As a consequence, they even earn more money. A person's level of optimism in the first year of law school predicted his income a decade later. An increase of one tiny point on the optimism scale was worth an extra $33,000 a year.[29]

Hope, whether internally generated or coming from an outside source, enables people to embrace their goals and stay committed to moving toward them. This behavior will eventually make the goal more likely to become a reality. It won the Lakers their repeat victory, Clever Hans his unique abilities, and Peter a long life following cardiac arrest. When our hopeful predictions turn out to be wrong, well, then we, like Bengtsson's participants, simply learn from our errors and try again. As the old saying goes, all's well that ends well; if it is not yet well, then it is not quite the end.

CHAPTER 4

What Do Barack Obama and Shirley Temple Have in Common?

When Private Optimism Meets Public Despair

A WAVE OF OPTIMISM SWEEPS THE NATION read the headlines. And sweep it did. Surveys reported 80 percent of Americans were optimistic about the next four years, 63 percent were confident that their personal financial situation would improve, 71 percent believed the economy would get better, and 65 percent said they thought the unemployment rate would drop.[1] Optimism was everywhere, and it did not remain in the "land of the free." Soon enough, hope reached the Spanish, Italians, Germans, French, and even the cynical Brits. From the United States and Europe, optimism steadily crossed more oceans, reaching India, Indonesia, Japan, Mexico, Nigeria, Russia, Turkey, Chile, China, Egypt, and Ghana, where, surveys reported, people were the most optimistic of all. Of the seventeen thousand people polled by the BBC and GlobeScan Incorporated, three out of four respondents expected positive change. A majority of fifteen out of seventeen countries, from East Asia, Latin America, and West Africa to the Islamic world, agreed that the near future would be, well, rosy.

What prompted such encompassing, unshakable hope? When were these glorious times? What events in human history could produce optimism in 80 percent of people in a nation of almost 304 million, and proceed to encourage people around the globe?

Take a wild guess. Was it during the bull market of the 1990s, when the U.S. market and many other global financial markets rose rapidly? Was it just after the fall of the Berlin Wall, when democracy seemed to have taken over and the Internet became accessible to people around the world? Did widespread optimism and joy follow the end of World War II, bringing a sense of relief to millions of people around the world? Or did hope climax following the first moon landing in 1969, when humans felt they had at long last conquered the world?

The answer is no, no, no, and . . . no. Optimism did not sweep the world in an unprecedented manner during times of financial stability, economical growth, scientific and technological achievements, or world peace. No, optimism rocketed during a period when we were deep in a global recession. The United States in particular was facing extremely challenging times, maybe one of the worst periods in its economic history, second only to the Great Depression of the early 1930s. In addition, the war on terror was in full swing, and many American soldiers were fighting in Iraq. It was late in 2008 when optimism peaked and an upbeat view was adopted by people around the world.

You may have already guessed what triggered hope in the American people during those dreary times. It was a forty-seven-year-old, born in Hawaii, father of two, author of *The Audacity of Hope*—Barack Obama, the first African American U.S. president.[2] How could one person inspire so many to become so hopeful? Significantly, not hopeful about their *personal* future (which we now know is fundamental to human nature), but hopeful about the future of a nation, the future of the world (which I will later argue is quite rare)?

David Gardner was one of many journalists who went out on the street looking for answers. He asked people why they thought Obama had managed to fill the country with hope. "[He] brings a different set of values, a different viewpoint to Washington, to the White House in particular, that we haven't seen there in the last eight years," said one woman. Another respondent attributed the nation's optimism to Obama's "openness to other ideas, his willingness to take advice from others."[3] Indeed, Obama's manifesto was built on change. As the first African American president, he represented progress, and not only for minorities. His ideology and plans for the nation promised developments that could lead to increased financial equality, market stability, better international relations, and more. All of this, together with his trustworthy demeanor, which was perceived by many to be in stark contrast to the demeanor of his predecessor ("He's not Bush," said one), could have led to the optimistic expectations of the people.

That was not the whole story, though. Would the people have been so hopeful, so optimistic, if the United States had not been in the midst of one of the worst financial crises it had ever known? Would the people of Egypt, China, Russia, and Ghana all have believed wholeheartedly that better things were just around the corner if hundreds of people had not been killed daily in military action? Ironically, the answer is no. As one interviewee put it, "[Obama] cannot not succeed. He has to succeed because the world really depends on him right now."[4]

High expectations regarding a new president are not limited to unstable times. Following the 2000 elections, 49 percent of Americans believed George W. Bush would be an above-average president. By the end of his first term, only 25 percent said he was.[5] When Tony Blair was elected Britain's prime minister, 60 percent of U.K. citizens believed the country was getting better. This figure fell to 40 percent by the time he left 10 Downing Street.[6] We may learn in the years to come that Obama was indeed worthy of the faith bestowed upon him. The fact, however, is that the people's urgent need for good news most likely drove positive expectations well above the baseline.

It is during hard times that people rely on optimism the most. When the going gets tough, we desperately start searching for the silver lining, and Barack Obama provided just that. In his inauguration speech, he fully acknowledged the challenging times:

> That we are in the midst of crisis is now well understood. Our nation is at war against a far-reaching network of violence and hatred. Our economy is badly weakened. . . . Homes have been lost, jobs shed, businesses shuttered. Our health care is too costly, our schools fail too many, and each day brings further evidence that the ways we use energy strengthen our adversaries and threaten our planet.[7]

However, in addition to the present hardships, he described what he believed would be a promising future:

> The road ahead will be long. Our climb will be steep. We may not get there in one year or even in one term, but America—I have never been more hopeful than I am tonight that we will get there. . . . America can change. Our union can be perfected. And what we have already achieved gives us hope for what we can and must achieve tomorrow.[8]

Putting Our Trust in Hope

If you were one of the millions around the world listening to Obama's victory speech on TV, on the radio, online, or in person, try to go back to that moment. How would you describe your emotional reaction? What were your feelings? "A feeling of spreading liquid warmth in the chest and a lump in the throat" was how a student at the University of California, Berkeley expressed his reaction. Does this come close? Scientists at Berkeley called this feeling "elevation."

Jonathan Haidt, a psychologist at the University of Virginia who studies the feeling of elevation, describes such instances as erasing cynicism and generating hope and optimism.[9] He has proposed a specific physiological mechanism that may underlie moments of elevation. According to Haidt, such occurrences stimulate the vagus nerve, which, in turn, triggers the release of oxytocin.[10] The vagus nerve is one of the twelve cranial nerves. Its course starts in the brain stem, which is an evolutionarily old part of the brain that plays a key role in regulating vital functions. The nerve extends all the way from the brain stem through the neck to the chest and abdomen. It conveys sensory information to the brain that reflects the body's internal state, as well as sends information from the brain to the rest of the body. Haidt suggests that stimulation of the vagus nerve by events that generate elevation causes the release of oxytocin.

Think about the last time you hugged someone, held a baby, patted a dog, or had sex. In all those instances, oxytocin was being released in your body. Oxytocin is produced in the hypothalamus (a structure in the base of the brain that produces neurohormones) and stored in the pituitary, which is situated just beneath the hypothalamus and secretes hormones. When triggered, oxytocin is discharged into the bloodstream, and it also binds to receptors in the brain, particularly in regions involved in emotional and social processing.

High levels of oxytocin reduce our uncertainty about social stimuli. A smile will be interpreted with more confidence as a positive signal, thereby reducing social anxiety and promoting approach behavior. In one brain-imaging study, volunteers were administered oxytocin before viewing faces conveying different emotional expressions. After the administration of oxytocin, the participants had less difficulty interpreting the expressions. As a result, less activity was observed in the amygdala, which is normally engaged in processing social signals, especially ambiguous ones.[11]

Reduced social stress and uncertainty, along with an increase in approach behavior, should enhance trust among individuals. Can administration of oxytocin, therefore, make people more trusting of strangers? To test this, scientists turned to their favorite method for investigating trust—a game known as, well, "the trust game."

Two participants play the game; one is the designated "investor" and the other is the "trustee." Let's say Average Joe is the investor and B. Madoff is the trustee. Both Joe and Madoff are endowed with an initial sum of money—twenty dollars. Joe, the investor, can send money to Madoff, the trustee. Let us say Joe decides to send Madoff five dollars. When Joe gives Madoff the money, the experimenter triples the amount. So now Madoff gets fifteen dollars. Madoff can then decide to keep all the money (in which case poor Joe just lost five dollars), or he can send a certain amount back to Joe. A fair move on Madoff's part would be to send back more than five dollars, maybe seven dollars. Joe has to sit back and trust in Madoff's sense of fairness.

When the game is played in the lab, scientists find that the Joes often choose to trust the Madoffs with their money.[12] The interesting bit is that if you squirt oxytocin up Joe's nose, he becomes even more likely to trust Madoff with his money. He shouldn't. Madoff does not reciprocate by sending greater amounts back.

Did Obama's public speaking trigger oxytocin release in his listeners? Given the overwhelming trust bestowed upon him, I would not be surprised if oxytocin levels were indeed heightened in the crowd. People trusted Obama. Obama was promising a better future, and so people put their faith in hope.

While during the first decade of the twenty-first century the

American people turned to their first African American president for hope, in the third decade of the previous century, they turned to a less likely source—a little girl with golden curls, a cheerful voice, and a reassuring baby face. Her name was Shirley Temple.

Shirley Temple, the famous child actress, appeared in numerous films during the 1930s. Her uplifting movies, in which she often sings and dances, made more money at the box office than those of any other star of her time—all this at the height of the Great Depression. The Great Depression originated in the United States following an abrupt collapse of the stock market in 1929, then evolved into a worldwide economic crisis, affecting cities all around the world. Personal income fell drastically; in the United States, thirteen million people joined the ranks of the unemployed and five thousand banks went out of business.[13] Just as in the credit crunch that would hit almost eighty years later, people sought someone to elevate their spirits. Into the void came Shirley.

Like Barack Obama's speeches, Shirley Temple's films mirrored a difficult era while at the same time promising a better future. Her movies offered a sense that better times were "just around the corner" (the title of a popular film she made in 1938) and that we should all just "stand up and cheer" (the title of a film she made in 1934). The fundamental premise of Shirley's movies was that everything would be okay as long as everyone worked in unison and cared for one another. Sound familiar? "It is the kindness to take in a stranger when the levees break, the selflessness of workers who would rather cut their hours than see a friend lose their job which sees us through our darkest hours . . . those values upon which our success depends, honesty and hard work, courage and fair play . . . —these things are old."[14] Akin to Obama's optimistic message, Temple's registered with the people. So much so that Franklin Roosevelt, who was the president at the time, said, "As long as our country has Shirley Temple, we will be all right."[15] Talk about optimism.

Global Pessimism

The reliance on optimism during the Great Depression, as well as during the recession of 2008, is intriguing not only because of the stark contrast between the difficulties of the times and the hope of the people but also because public optimism is relatively rare. While private optimism (positive expectations regarding our *own* future) is commonplace, it is typically accompanied by public despair (negative expectations regarding the future of our country). Statistics such as those that conveyed people's optimism about the economy in 2008 (71 percent believed the economy would get better in the upcoming year)[16] and the political situation (three out of four respondents expected positive change in international relations) are not frequently observed during stable times. When public optimism is observed, it is usually short-lived (such as hope being elevated shortly after elections).[17]

More often than not, people expect to do better personally in the near future while anticipating that the rest of the country will go down the drain. For example, a few months before the financial collapse of 2008, the majority of British people said they thought the condition of the country was getting worse—they would soon discover they were correct. At the very same time, they expected their personal circumstances in the coming years to improve—many of them would find they were wrong.[18] Ninety-three percent said they were optimistic about the future of their *own* family, but only 17 percent were optimistic about the future of *other* families.[19] That year, most people expressed satisfaction with their personal experience with Great Britain's National Health Services (NHS): Almost 80 percent were satisfied with their last visit to the hospital, and 65 percent said their *local* NHS was doing a good job. However, at the same time, a majority of respondents said the NHS was in crisis, and less than 50 percent said the NHS was offering good service *nationally*.[20]

It was not only the National Health Services that the British were pessimistic about. They were not so positive about the prospect of the government controlling crime and violence, either.[21] The vast majority (83 percent) believed that violence was rising in the United Kingdom. In reality, crime had been going down steadily for nearly a decade. This was partially because national spending on crime prevention was rising. Nevertheless, people's confidence in the government's ability to stop crime went down from 63 percent in 1997 to 27 percent in 2007. Crime seemed to be a major concern for citizens. More than half of the respondents rated crime as one of the three most worrying issues in Great Britain. Citizens of other countries did not seem so worried. Only 23 percent of Spaniards and 35 percent of Americans listed crime as one of the three most troubling issues in their country. Did the Brits have special reason to be concerned? Should they all have moved to the United States? Maybe, but crime should not have been their reason.

Let's take a quick look at homicide rates in countries around the world. Below are the most recent figures for homicide rates per 100,000 population in the countries listed:*

El Salvador—71
Guatemala—52
Colombia—35
Brazil—22
Mexico—15

*Some of the homicide rates below include attempts and some do not. This limits comparison between countries. Those rates that exclude attempts have been starred.

Russia—15
Thailand—5.9
United States—5
France—1.31*
Australia—1.3*
United Kingdom—1.28*
Italy—1.1*
United Arab Emirates—0.92
Japan—0.5[22]

In other words, if a Brit decides to pack his things and move to France, or even Australia, he is more likely to end up a victim of homicide. While countries in South America and Central America head the list of places with high crime rates, Italy, which is known for its Mafia, seems to be fairly peaceful. Although the United Kingdom is not doing as well as Italy, homicide rates are still relatively low. Granted, homicide is only one type of violent act, but it is not a bad indicator of general crime rates.

It seems the British might be overestimating the extent to which crime is a problem. Does that mean they also overestimate their *own* likelihood of being victims of crime? If acts of crime are one of the top three disturbing issues in the United Kingdom, and people believe the ability to put a stop to it all is decreasing, does it not follow that one should be quite worried regarding one's *own* safety? My student Christoph Korn and I decided to investigate. We gathered data regarding the probability of falling victim to different crimes, then asked Londoners to estimate their likelihood of encountering such incidents in order to see if they were on the mark.

The experiment, which we conducted in our lab in central London, indicated that people slightly underestimated the probability of being victims of crime.[23] When asked to estimate the likelihood of different adverse events happening to them in their lifetime, such as having a car stolen, being mugged, having their apartment burgled, and other traumatic events, people gave estimates that were on average a bit lower than the figures published by the authorities. Thus, while people believe crime rates are high, they also believe *they* are somehow immune. While the economy of the country is in trouble, we believe *we* will endure. While health services are poor, and public schools are even worse, *our* local services and local schools are, fortunately, quite good.

In a symposium organized by the Royal Society of Arts in 2008, politicians, academics, and polling experts came together to discuss the contradiction between private optimism and public despair. At the end of the session, a man in the audience raised his hand and said, "I do not see what the big fuss is about. It is true. The polls are right. National Health Services is a mess, and in general provide poor service. However, my own branch is indeed excellent."[24]

Switching Between Optimism and Pessimism

Why do we see such a disconnect? Why is it that people continually underestimate their own risks while overestimating the severity of the situation for the rest of society? Why do we rate our own experiences highly while at the same time assuming that services in the rest of the country are poor? If we will survive the economic downturn and will avoid being mugged, surely won't everyone else?

Deborah Mattinson, who spoke at the Royal Society of Arts symposium, suggested it is all about a subjective sense of control. People tend to feel more optimistic about things they believe they can control. Often, this sense of control is just an illusion.

Nevertheless, when we think our fate is in our own hands, we are more likely to believe we can steer the wheel in the right direction. We believe we are less likely to be mugged than the next person because we will surely avoid walking in dark alleys. We are less likely to have skin cancer because we will not expose ourselves to the sun unprotected. We will survive the economic breakdown because our unique talents are always in high demand. At the same time, we acknowledge that we have no control over the financial situation of our country or over the health and safety of our fellow citizens. Thus, we are usually less confident that those elements are heading in the right direction.

There is one additional factor that comes into play—the power of relativity. Whether we are satisfied with our salary, dinner order, cell-phone carrier, and health services depends to a large extent on how much our friends are making, whether our dinner partner got a more appetizing entrée, how much our colleague is paying for his monthly mobile plan, and whether our family doctor is more competent than the other practitioners. Believing that our own positive experiences do not apply to the general population means we are actually privileged. It is not just good services we receive; it is *superior* services. If we think our local public school is excellent while the rest of the schools are inadequate, that means we lucked out. Not only do our kids get a great education; they receive a better education than all the other kids in other public schools around the country.

Once again, the brain plays a little trick that boosts positive illusions. Not only do people hold an optimism bias about their personal future; in addition, they hold a pessimistic bias about everyone else's. Put one and one together, and instead of merely believing we are fortunate, we perceive ourselves as exceptionally fortunate, which is even better. And when we encounter troubled times, it is always helpful to believe that everyone else is doing badly, too.

At the end of the day, our expectation regarding the future of our society will be positive or negative depending on which view feeds private optimism best. During tranquil times, public pessimism may fuel private optimism by way of comparison. So when the world is doing okay, having a pessimistic view of society while maintaining optimistic beliefs about our own future means that not only do we expect to do better; we expect to do better while others do badly. This gives us an illusion of superiority. It does not mean we are ill-wishers; it merely suggests that the rosy spectacles we use to view our future are not worn when examining the future of our fellow citizens. In fact, we often use dark shades to assess the future of our country.

However, when society reaches unprecedented lows that affect our personal lives directly, the only way for our situation to improve is to take the rest of the world upward with us. When people lose their jobs and savings during times of extreme economic distress, they have to believe that the world will soon take a turn for the better, because that is the most likely way for them to be able to regain income and well-being. That is the time when people turn to bearers of good news such as Barack Obama and Shirley Temple. That is when optimism sweeps the world. Or at least it does until the economy stabilizes, at which point we are quite happy to go back to public pessimism.

CHAPTER 5

Can You Predict What Will Make You Happy?

The Unexpected Ingredient for Well-being

What do you think will make you happy? Jot down five things you believe will increase your satisfaction with life. Will earning more money make you happy? How about exercising more? Or spending more time in the sun? A survey of 2,015 people conducted by the British research company Ipsos MORI revealed that people believe the following five factors are most likely to enhance their happiness (they are listed in order of importance).

1. More time with family
2. Earning double what I do now
3. Better health
4. More time with friends
5. More traveling[1]

Were any of these on your list? The answer may depend on how old you are. The perceived importance of the above factors varies depending on age. The significance of having more money, for example, steadily decreased as a person aged. While 55 percent of young adults, ages fifteen to twenty-four years, believed being richer would make them happier, only 5 percent of respondents seventy-five years and older thought more money would contribute to their happiness. Maybe life experience had taught them that happiness cannot be bought. On the other hand, the perceived contribution of health to satisfaction with life increased steadily as a person aged. Only 10 percent of respondents from fifteen to twenty-four years old rated better health as one of the top five factors that would make them happy, as opposed to 45 percent of people over the age of seventy-five. This is not very surprising; older adults face far more health problems than younger adults and thus are more likely to be concerned with their physical condition. The perceived contribution of spending more time with family to people's well-being remained relatively constant with age, although it peaked between the ages of thirty-five and forty-four. This may reflect the anxiety felt by people of that age group with regard to adequately balancing a professional life with a family life.

Could we spend more time with our family and friends, travel more, and at the same time proceed to earn double what we do now? This would be challenging for sure. And if we did manage to land a high-paying job that brought in the dough as well as left us enough time to take our family and friends for a vacation in the Bahamas, would that make us happy? Could there be other things within our immediate reach that would make us just as happy? And ultimately, are we able to predict what will make us happy?

Estimating what will enhance our well-being is not an easy task. Most people are more than willing to tell you what will make you happy. Advertisements try to sell us joy in a soda can or a chocolate wrapper. Our society preaches that education, marriage, children, and money will make us happy. Faith, sex, world peace, drugs, love, owning a house, a job on Wall Street, a retirement plan, ice cream, cable TV—what really matters?

Consider the following list of factors. Some of them are positively associated with individual reports of satisfaction with life, some of them are negatively related, and yet the relationship of other factors to such satisfaction is so complex that certain surveys classify them in one group and certain surveys in the other. Can you guess which ones are which?

1. Walking / swimming / playing sports
2. Marriage
3. Having children
4. Gardening
5. Going to church / synagogue / mosque / another religious center
6. Having a Ph.D. or another degree
7. Having a high income

So, do you think you know what makes people happy? Let's start by reviewing the factors that were positively associated with life satisfaction. Ben Page, managing director of Ipsos MORI, who conducts global surveys to figure this out, summarized his findings by quoting a traditional saying: "If you want to be happy for a few hours, get drunk. If you want to be happy for a few years, get a wife. If you want to be happy forever, get a garden." Could there be any truth in this axiom? Grab your bucket and spade and start getting muddy. Page's survey of thousands of individuals reveals that people who garden at least once a week are happier on average than the rest of us, and people who never garden are less likely to be satisfied with their life.

You do not even have to garden per se. Studies show that just taking care of a plant is positively related to well-being. According to scientists at Texas State University, employees who kept a plant in their office were happier than those who did not.[2] This does not necessarily mean that having a plant or spending time gardening once a week *leads* to life satisfaction. Neither does it imply that happiness makes people grab a shovel and start planting roses. Although all these relationships are feasible, it is also possible that gardening and happiness are correlated because they are both caused by a third factor. For example, having more free time may increase both well-being and the likelihood of working in the garden. Having a good relationship with your coworkers may contribute both to your satisfaction with life and to your desire to make the office welcoming by decorating it with a plant. So although we cannot conclude that plants *make* us happy, we can presume that a coworker who has a leafy one on his desk is probably happier than a colleague whose desk is piled with papers sans greenery.

What else is related to happiness? According to the survey, if you hold a Ph.D., go to church (or another religious center), and play sports, you are several times more likely to be happy-go-lucky than someone who does not have a Ph.D., never goes to church, and shuns physical activity. Thirty-five percent of the respondents who held a higher degree described themselves as very happy, while only 23 percent of people without such a degree did. Almost half of the respondents who frequented their church several times a week reported being very happy, compared with only 26 percent of people who never did. Again, these numbers convey only a *correlation* between the factors, not a *causal* relationship.

Married with Children

How about the conventional wisdom that having children will make us happy? Most people believe that having kids is crucial for leading a satisfactory life. People spend an incredible amount of time, effort, and resources on their kids. Do our kids make us happy? Studies consistently show that if there is any correlation between having offspring and happiness, it is a negative one. For example, data from Ipsos MORI shows that satisfaction with life decreases steadily once a married couple who have no children begin raising a family, and reaches rock bottom when the kids are teenagers. From that time on, happiness gradually increases, and goes back to prechild levels of happiness when the nest is empty. In fact, middle-aged people (thirty-five to fifty-four years old) are the least satisfied. Subjective well-being is highest in young adults (fifteen to twenty-four years old), with older adults (seventy-five years and older) coming in at second place.[3]

The notion that raising children is negatively related to happiness was supported by a study conducted by the Nobel Prize winner Daniel Kahneman.[4] Kahneman, a cognitive psychologist, received the Nobel Prize in economics in 2002 for his work on prospect theory, which describes how people make decisions under uncertainty. He is known for his research in behavioral economics and decision making, and is especially recognized for describing cognitive biases—the basis for many human errors. In his later years, Kahneman has focused on studying hedonic psychology. Using principles developed in his earlier work, he now attempts to describe and explain the errors we make in thinking about what will make us happy.

In one study, Kahneman and his team investigated life satisfaction in a large group of American and French working females. To measure happiness, they used a relatively unconventional method. Instead of questioning people regarding their general well-being (i.e., simply asking them how happy they were with their life), they asked people to report their emotional state on a moment-to-moment basis. The participants were requested to pause throughout the day and state their present feelings as well as their activities. This technique produced what Kahneman calls "a measure of experienced happiness." According to Kahneman and his colleagues, this is a more accurate measure of happiness than that derived from traditional techniques. The idea behind the measure of experienced happiness is that well-being is most significantly influenced by the flow of our daily experience. What really matters is when, and how often, we feel irritated, anxious, satisfied. Although we do at times look back on our lives and assess our existence, we do not do so frequently. Our happiness is thus not affected to a large extent by reflecting on our lives, but by the flood of feelings that are constantly generated within us. Most questionnaires of subjective well-being, however, ask us to reflect and assess our general satisfaction with life rather than our daily experienced happiness.

Using the measure of experienced happiness, Kahneman and his colleagues found that the experienced happiness of mothers was negatively correlated with the amount of time they spent with their kids. The participants reported experiencing fewer moments of joy while interacting with their children than while performing household chores such as cooking and grocery shopping. In fact, there were not many other activities that had less of a positive influence on experienced happiness—except for a daily commute. Commuting to and from work contributed most negatively to satisfaction with life.

The fact that being stuck in traffic for hours or in a steaming-hot subway car, trapped between hundreds of people trying to get home during rush hour, does nothing to enhance our well-being is not shocking. The fact that playing with our kids, reading to them, feeding them, or going over their homework does not do much more is startling. It is also disturbing. The consistent conclusion across studies that children do not necessarily bring us joy conflicts with conventional wisdom in a striking manner. Why is it that popular culture and the people around us would have us think otherwise? Why do individuals insist, and often strongly believe, that their happiness is rooted in the existence of their offspring? One explanation is that happiness, whether experienced or reflected, is not necessarily the most significant factor for the continuation of humankind. Passing on our genes, on the other hand, is. With that adaptive goal in mind, it seems plausible that while rational scientific methods suggest raising children does not make us happy, people will be biased to believe otherwise. The Harvard psychologist Daniel Gilbert suggests that people rationalize the effort and resources they devote to child rearing in terms of the happiness their children cause.[5] Because our society often views happiness as an ultimate goal, we conclude that we must be spending so much of our time and energy on our children because of the joy they bring us, rather than because of a biological urge imprinted in us to pass on our genes.

One may try to explain in similar terms the popular belief that marriage brings happiness. Whether the relationship between marriage and happiness is a myth or not is an unresolved question. In contrast to other factors, such as changes in health or wealth, marriage can affect well-being in both a positive and a negative direction. For certain individuals, marriage increases happiness, while for others, it decreases it; thus, on average, it is difficult to detect an effect. A study conducted in Germany found that marriage did not influence happiness much. Well-being increased somewhat after marriage but then quickly returned to previous levels.[6] Interestingly, people who do show long-term benefits from marriage are individuals who started out with relatively low levels of happiness. People who were already happy before marriage do not gain much from tying the knot. This may be because people who enjoy high levels of well-being are more likely to have a fulfilling job and a close circle of family and friends, and thus have relatively less to gain from marriage.

A survey by Ipsos MORI found that 33 percent of married individuals described themselves as very happy, 31 percent of individuals cohabiting did, and only 25 percent of singles did. These figures may indicate that what impacts our well-being is not necessarily having a marriage certificate, but, rather, feeling loved and having the subjective feeling of security that often comes from sharing a life with another person.

More, More, More

The factor that is most debated in the literature with regard to its relationship to happiness is wealth. Would we be happier if we were richer? According to the Pew Research Center, people who earn more are happier on average.[7] While half of the individuals surveyed who earned $100,000 or more a year said they were happy with their life, only 25 percent of the respondents who earned $30,000 or less a year declared they were happy. However, data from different sources, including the U.S. General Social Survey, shows that while there is indeed a twofold increase in subjective reports of happiness between the group with the lowest income and the group with the highest income, the difference between people with median-range income and top earners is not large.[8] Similarly, when examining happiness levels across different countries, a steady increase in well-being is observed as a function of gross domestic product (GDP). However, once GDP reaches a certain level, an additional increase is no longer correlated with greater subjective well-being.[9] According to figures published by Ipsos MORI, Britain's rising GDP over the last fifty years has done nothing for citizens' reported happiness.

Studies that use a measure of moment-to-moment experienced happiness to quantify well-being do not find a clear relationship between income and happiness. In a study where participants were asked to report their feelings every twenty-five minutes throughout the day, the correlation between moments of happiness and income level was zero (no relationship at all). It was revealed, however, that people who earned more tended to report more moments of anger, anxiety, and excitement, suggesting that high-paying jobs may produce higher levels of arousal, which are more often than not of the negative kind.

These findings raise an obvious question: How come we strive to earn more if having a larger bank account does not make us happy? There are a few possible explanations. First, while we crave more, as with almost anything in life, once we get what we thought we wanted, we quickly adapt and it no longer enhances our pleasure. Even lottery winners, merely a year after winning the lottery, report similar levels of satisfaction with life as they did before winning millions.[10] We adapt to our newly acquired goods—our large-screen TV, the BMW outside our five-bedroom house, and our designer clothing. We may desire to own a new home, have a fancier car, go on vacation more often, eat at high-end restaurants, and buy expensive suits. However, once we have all of this, within months we acclimate and the extra cash no longer contributes significantly to our level of happiness.

Moreover, a higher income usually goes hand in hand with greater responsibilities in the workplace and longer working hours. Recall that when asked what would make them happy, people rated spending more time with family as number one, earning more as number two, and spending more time with friends as number four. In order to get ahead professionally and earn a fatter paycheck, we are often required to sacrifice time with family and friends. We might be attaining goal number two, but to do so we are sacrificing goals number one and number four. As a result, we may well be increasing our overall feeling of achievement, while at the same time decreasing the amount of time a day that we feel happy.

Such dissociation between the influence of wealth on our general feeling of satisfaction, on the one hand, and moment-to-moment pleasure, on the other, may explain why studies find different degrees of correlation between income and subjective well-being as a function of the surveying method. While studies using general questions to assess happiness (such as "How happy are you in life?") find significant mid-range correlations between income and happiness,[11] studies using momentary experienced happiness often do not find a significant relationship at all.[12] This is because a higher income may indeed influence reflected satisfaction with life without significantly enhancing our experienced happiness.

Perspective

Another reason for the above dissociation is what Kahneman and others term the *focusing illusion.*[13] This is the exaggerated importance we attribute to specific aspects of our lives when we are asked about them. A survey that inquires about our income will bias our attention toward our financial situation. Thus, when later answering general questions about satisfaction with life, we are likely to place greater weight on economic considerations than we normally would. For example, if people are asked first about their income and then about satisfaction with life, a greater correlation is revealed between the two than if the order is reversed. As Kahneman notes, "Nothing in life is as important as when we think about it."

Several researchers have suggested that what contributes to our happiness is not our absolute wealth, but, rather, our *relative* wealth—that is, our material means relative to that of the people around us.[14] We may be happy earning $80,000 a year if our neighbors, peers, and relatives are all earning about $50,000. However, if our coworkers and friends were earning $95,000 annually, we would no longer be satisfied with only $80,000 a year. The importance of relative wealth to our well-being explains counterintuitive findings in the literature, such as why an increase in a nation's GDP over time is not accompanied by an increase in general subjective well-being. While the country may be getting wealthier, the individual's relative economic status remains the same, and thus happiness stays constant.

Relativity is a crucial aspect of human psychology. Consider the way we perceive our physical environment. The extent of change that needs to occur in the state of our surroundings in order for us to notice a difference (such as a change in brightness or loudness) is dependent on preliminary states. For instance, if we are listening to music at a low volume, a small change to the volume will easily be noticeable. However, if we are listening to our favorite band at full volume, a larger change needs to be induced in order for us to detect a difference in loudness. The importance of relativity to perception was initially pointed out by Ernst Heinrich Weber, a German physician who is sometimes thought of as the founder of experimental psychology. In one of his experiments, Weber gave blindfolded volunteers weights to hold. He slowly increased the amount of weight and asked his participants to tell him each time they felt a difference in heaviness. Weber found that the increase in weight needed for a difference to be perceived was proportional to the starting value of the weight. So while a few additional ounces were enough to induce a noticeable difference at first, as the weight increased, a few ounces were no longer sufficient, and a whole pound was required to induce a sense of change.

People perceive money in a similar fashion. If you give someone ten thousand dollars to do a job and then add fifty dollars as a bonus, the person will not be that impressed. However, if you pay someone seventy dollars to do a job and then add another fifty dollars, the person will be extremely thankful. Studies in behavioral economics show that the value people assign to money is best described in a nonlinear fashion.[15] If the money earned for a job is doubled, people do not perceive the value of that job as double, but as a bit less than double. Like the perception of loudness or light, the subjective value of a dollar depends on the starting point. A dollar is valued more if you start off with only one dollar than if you start off with one hundred dollars. Therefore, the more you have, the more you need to increase your wealth in order to even notice a difference that would affect your happiness.

While there does not seem to be a clear relationship between wealth and subjective well-being, people strongly believe that such a relationship exists. A study published in the journal *Science* reported that when participants were asked to estimate the mood of other individuals, they predicted that people who earned more were also more likely to be in a better mood.[16] In reality, this was not the case. So although people seem to think wealth, children, and marriage will make them happy, hard science suggests otherwise. Remember, even winning millions in the lottery does not seem to make us feel better for more than a few months. Yet every week people spend well-earned money on lottery tickets, hoping to become richer and thus happier with life.

Is Our Sense of the Past Just as Hazy as Our Sense of the Future?

Reflect back on your childhood years. Which events come to mind? Many people will recall a birthday party, punishment for a bad deed, a school play, a sports game, a fight with a friend, a school outing, or a childhood love. These memories survive the test of time because they elicit strong emotional reactions. They are the ones that remain vivid in our minds and are easily retrieved. Memories of everyday events, such as a history class or going shopping, are just not as distinct and tend to fade away.

In 2004, I spent ten months at the University of California at Davis conducting experiments that were designed to demonstrate the effects of emotion on memory. Davis is a small town in northern California, about a ninety-minute drive from San Francisco and forty minutes from the beautiful Napa Valley wine country. It is a quiet, peaceful town, with one main street, a few restaurants, wide lawns, courteous residents, and warm weather year-round. Needless to say, having arrived from New York City, I was in shock. The slow-moving atmosphere and the friendliness of the locals were entirely alien. My landlady left candy on my doorstep on Halloween, Christmas, and Easter. She did not ask to see the numerous documents I had put together from my employer and my bank, or other references indicating that I was sufficiently trustworthy to rent her apartment. I didn't need my earplugs to sleep undisturbed—a previous necessity for shut-eye on Sixth Avenue—and there was nowhere to go after 10:00 p.m.

The sleepy town, however, had a lasting effect on my research, thanks to the memory expert Andrew Yonelinas, a psychologist renowned for his dual-process theory of recognition. According to dual-process theory, memory retrieval includes two distinct processes: familiarity and recollection. Imagine you are walking down the street when suddenly someone stops to say hello. You look at the individual and feel you have encountered him before. The guy is *familiar;* you know you have previously met, but you're not sure of the circumstances. Knowing whether someone is new to you (never seen before) or old (previously encountered) is based on a sense of familiarity. We can feel someone is familiar without necessarily remembering when or where we have come across the person before. You politely engage in conversation with the stranger while feeling somewhat awkward, as you are unsure what exactly the two of you have in common. Once the person in front of you mentions Sally, a good friend, you suddenly recall meeting a few months ago at Sally's dinner party. Remembering the episodic context (the dinner party) where you encountered Bob (eventually, you recall his name) is termed *recollection*—the ability to go back in time and mentally reexperience an event.

These two memory processes (familiarity and recollection) have been shown to be functionally and neuroanatomically distinct.[17] They rely on different brain regions within the medial part of the temporal lobes. While a region called the hippocampus is crucial for recollection but not for familiarity (the hippocampus is discussed in detail in chapter 2), the adjacent perirhinal cortex signals familiarity. Amnesiacs who have sustained damage to the hippocampus but whose surrounding cortex is intact usually suffer impaired recollection, but they know when something is familiar. They may know they have met you before, but they will have no memory of the episodic context of the encounter.

Emotion strongly enhances our recollective experience. It increases the confidence that we are remembering the incident as it actually unfolded, and the vividness of the image.[18] However, this enhancement in recollection does not happen at once; it takes time. In a study I conducted with Yonelinas at Davis, we presented volunteers with highly arousing emotional photos (mostly unpleasant photos of mutilated bodies and acts of violence) as well as neutral photos (people reading in a bookstore or employees working in an office). We then tested the volunteers' memory of half the photos immediately after presenting them; we tested their memories of the rest of the photos twenty-four hours later. At first, it seemed that the volunteers' memories of the emotional and neutral photos were not different; they remembered them equally well. However, when they came back to the lab a day later, something had changed. Now their recollection of the emotional photos was better than that of the neutral photos. The volunteers' memories were not always more accurate, but they reported they were more vivid.[19]

Having extremely vivid memories of past emotional experiences and only weak memories of past everyday events means we maintain a biased perception of the past. We tend to view the past as a concentrated time line of emotionally exciting events. We remember the arousing aspects of an episode and forget the boring bits. A summer vacation will be recalled for its highlights, and the less exciting parts will fade away with time, eventually to be forgotten forever. As a result, when we estimate how our next summer vacation will make us feel, we overestimate the positive. An imprecise picture of the past is one reason for our inaccurate forecasts of the future.

The two other principal factors that lead us to mispredict what will make us happy are the same ones that make us mispredict what will devastate us. The first is our tendency to underestimate our rapid adaptation to almost any new circumstance. Yes, a higher salary or better health may make us temporarily happy. However, sooner rather than later we will become accustomed to our large bank account and stronger physical state and will slide back down to our normal level of well-being. The thing is, we do not incorporate such adaptation into our forecasts, and therefore we inevitably make inaccurate predictions. Second, when we think about how a higher income, more vacation time, or better health will affect our happiness, we tend to focus on that one factor and disregard everything else, which, alas, will stay the same. We might have more cash in our wallets, but we will still need to commute to work every day and do the dishes. So while certain life changes may make us happier than we are now, they might not have as great an impact as we think they will.

This does not mean that change is impossible. Although happiness levels are relatively stable throughout life, modifications do occur. For example, a study conducted in Germany found that a quarter of the people surveyed reported significant changes in their satisfaction with life over a seventeen-year period.[20] We can, in principle, become more or less happy than we are at the moment. However, it is not what we think will matter that actually makes a difference.

What Really Matters

At the beginning of this chapter, I asked you to list five factors that would make you happy. If you are like most people, your list included more money, better health, or more time to travel. I am guessing your list did not include greater political stability. You might want to revise your list. Political stability is one of the nine strongest indicators of a nation's well-being, and human rights is one of the two strongest.[21] Other factors include national divorce rates and life expectancy at birth. Performing random acts of kindness one day a week has also been shown to increase happiness.[22] I bet none of you listed "Being kinder" between "Earning double what I do now" and "More traveling."

We are unable to predict accurately what will make us happy, but does that matter? We seem to be doing quite well. Although we are not great at guessing what will make us cheerful and content, most of us are quite happy. A large-scale survey that involved people in multiple countries reached the definite conclusion that the majority of people are happy most of the time.[23] *Eighty* (!) percent of the respondents said they were happy. From the Amish to the inhabitants of the Sahara Desert, many of us are purely and utterly joyful. What is the crucial factor that makes us so happy? All demographic factors put together can explain only 20 percent of the variance in happiness among individuals.[24] It is not health or attractiveness, wealth or marriage. Could it be that our *expectation* that a larger income, a healthier body, or a loving family will make us happy actually makes us happy?

Winning the lottery might not make us happy. However, buying a ticket in the belief that if we win, we will forever be ecstatic makes us skip with joy. The mere thought of what we will be able to do with all those millions momentarily fills us with a warm, fuzzy feeling. Our belief that happiness is just around the corner is, ironically enough, what keeps our spirits high in the present. Imagining a better future—which is attainable if we follow certain rules (or so we think)—maintains our well-being.

So what exactly happens in our brains when we imagine ourselves becoming the CEO of a company or graduating at the top of the class? In a study I conducted a few years back with Elizabeth Phelps, a prominent neuroscientist who was my thesis adviser, and students Candace Raio and Alison Riccardi, I asked volunteers to imagine specific events that might occur in the next five years while recording their brain activity with an fMRI scanner.[25] Some of the events were desirable (an enjoyable date, winning a large sum of money) and some were undesirable (losing a wallet, the end of a romantic relationship). The volunteers reported that their images of sought-after events were richer and more vivid than those of unwanted events. When they visualized a scenario in which they lost a large amount of money or broke up with their partner, only blurry images were created. However, when imagining an award ceremony, a detailed story emerged. How does the brain generate this bias?

Delilah, an energetic psychology undergraduate with curly blond hair and big eyes, was the optimistic type. When she imagined her graduation day, enhanced activity was observed in two critical regions of her brain: the amygdala—that small structure sited deep in the brain that is central to the processing of emotion; and the rostral anterior cingulate cortex (rACC)—an area in the frontal cortex that modulates activity in regions that are important for emotion and motivation. The rACC was assuming the role of a traffic conductor, enhancing the flow of activity from subcortical regions when those conveyed positive emotions and associations. The result was a powerful, detailed image: Delilah wearing her purple-and-black gown and her cap, holding an NYU diploma in hand while her family cheered in the background. The more optimistic a person is (according to standard psychological tests), the greater activity in these regions when he or she imagines positive future events relative to negative ones.[26]

These observations revealed an important biological link—a connection between optimism and depression. Rollo May, the American existential psychologist, said that depression is the inability to construct a future. As a matter of fact, clinically depressed individuals find it difficult to create detailed images of future events, and when they do, they tend to be pessimistic about them.[27] Two brain regions have been identified as being particularly malfunctional in cases of depression,[28] and the way these two regions communicate with each other is specifically abnormal. These structures are the amygdala and the rACC. The same neural pathways that break down in depressed patients are the ones that play a role in mediating the optimism bias in healthy individuals.

What we were observing in the brains of healthy, optimistic volunteers turned out to be a mirror image of the pattern of activity often detected in the brains of clinically depressed individuals. In depressed patients, the rACC fails in regulating amygdala function adequately. As a result, while healthy people are biased toward a positive future, depressed individuals perceive possible misfortunes a bit too clearly.[29] While severely depressed patients are pessimistic, mildly depressed people are actually pretty good at predicting what may happen to them in the near future—a phenomenon known as *depressive realism.* If you ask mildly depressed individuals what they expect in the upcoming month, they will give you a pretty accurate account. If you ask them about their longevity or the likelihood of having a certain illness, they will give you correct estimations. Could it be that without an optimism bias, we would all be mildly depressed?

The optimism bias is a crucial ingredient for keeping us happy. When people perceive the future accurately, when they are well aware that none of the things people assume will make them happy is likely to have any lasting significance on their well-being, when they take off their rose-tinted glasses and see things more clearly, they become depressed—clinically depressed.

CHAPTER 6

Crocuses Popping Up Through the Snow?

When Things Go Wrong: Depression, Interpretation, and Genes

Consider the story of two young men, Shawn and Fred. Shawn lives in Seattle with his girlfriend, Phoebe, and their dog, Mr. Kat. Fred lives about three thousand miles away in a condo in Florida with his wife, Sabrina. Life is pleasant for Shawn and Fred—both are successful corporate lawyers, healthy and happy in their relationships. One day early in October, the two fly out to Paris for a business meeting. This is not an unusual event; Fred and Shawn travel often. A week later, they return home, excited to reunite with their loved ones. When Shawn enters his lakeside house, however, something feels different. He soon notices things are missing—Phoebe's wardrobe is wide open and her clothes are gone. He frantically walks around the house, only to find that her books, her shoes, her DVDs, and her camera are also missing. Everything's gone, except for Mr. Kat, who is on the sofa, looking sad and confused. By way of coincidence, Fred arrives at his condo and finds a similar scenario. On the sofa, instead of Mr. Kat, he discovers a letter from Sabrina—the type of farewell letter you often find in soppy films.

Needless to say, both men are distraught. The people they cared for most have left, and they have done so in a way these two men previously imagined took place only on daytime TV. For the next couple of weeks, they are enveloped in their misery. They find it difficult to eat, sleep, or work, and they lose interest in social interaction and physical activity. They stay in bed, going over the events again and again in their minds, ruminating about what they did wrong and what they could have done differently. What would have happened if they'd never left for Paris? The thoughts spiral in their minds in an exhausting manner.

Shawn's and Fred's reactions are normal. People find loss painful. We also find failure, rejection, abandonment, and change difficult to deal with. Shawn and Fred experienced all of the above, and their situations were made particularly distressing by the fact that they had no apparent control over the outcome. Temporary sadness, inaction, and even hopelessness are to be expected in such cases. The behaviors they exhibit—loss of interest in previously rewarding activities, sleep disruption, weight loss, difficulty concentrating, low mood, and negative thoughts—are all symptoms of depression.[1]

Some psychologists argue that these reactions have an adaptive function.[2] We temporarily retreat into our own world, focusing our mental energies on processing the painful events so we can heal. We examine our behavior, the actions of others, and the circumstances that led to the specific outcome until we can make sense of it all. It is a time-out on life—similar to the rest required when fighting a cold. When we catch a cold, most of us stay in bed for a day or two, drinking tea and chicken soup, until our immune system effectively fights it off. Some, however, will develop complications. People who are prone to develop severe problems from the common cold are individuals with weak immune systems, such as the elderly, pregnant women, and people with preexisting medical conditions. The rest will successfully fight it off and return to normal health. Similarly, all of us will experience loss and adversity at some point, and most will initially react like Shawn and Fred. However, while the majority of people will eventually heal from disappointment and heartache, occasionally adversity will trigger a prolonged negative mental state, which may be the beginning of clinical depression. This will occur in approximately 15 percent of the population over their lifetime.[3] Often, depressive episodes can be traced back to a specific stressful event in a person's life, but not always.

After a period of processing and rumination, will Shawn and Fred become stronger or weaker? What will they have learned? How will they perceive their past, their future, and their role in what occurred? Here, the two men diverge. Fred blames himself for what happened; he concludes that he was demanding and inflexible in his relationship with Sabrina. He begins to believe that his high expectations will forever interfere with his relationships. He thinks his life is doomed and that his inability to compromise not only will destroy his romantic relationships but also will make him a poor lawyer and, when the time comes, an inadequate father. Fred's manner of interpreting the event is known as a *pessimistic explanatory style.* He holds himself responsible for the negative event ("I drove her away"), he believes the situation is permanent ("All my relationships are doomed and there is nothing I can do about it"), and he extrapolates from a specific failure to other life domains ("I am not only a terrible partner but a poor lawyer and friend").[4]

Fred thinks negative events are inevitable in his life because of what he considers to be problematic personality traits. His current inability to mend his relationship with Sabrina grows into a general hopelessness about the future. Fred supposes that the failure of his marriage indicates he will fail in all future romances. He becomes pessimistic, expecting the worst.

Shawn has a very different interpretation of the situation. Yes, he and Phoebe did not always see eye to eye. Yes, he has made mistakes, but he is human—who doesn't stumble once in a while? Ultimately, it was Phoebe's weaknesses, her inability to deal with conflict, her insecurities, that made her run away rather than deal with reality. He will need to seek a stronger, more trustworthy partner. Shawn uses what is known as an *optimistic explanatory style.*[5] He sees others as causing the adverse situation ("Phoebe is weak and hysterical"), believes his circumstances will change ("I will find another partner"), and does not globalize the failure in his personal life to other domains ("I am still a successful lawyer"). Because Shawn views Phoebe's departure as an isolated event, one that has no bearing on future relationships (or other social and professional interactions), he is hopeful. He may have been unable to stop Phoebe from leaving, but that does not mean he has no control over future relationships. To the contrary, he believes he has learned an important lesson. He is optimistic that as long as he steers clear of the Phoebes of the world, he will be fine.

Of course, neither Fred's nor Shawn's interpretation of the event is fully aligned with reality. Most likely, both had some role in the breakdown of their relationships. Probably both will go on to make some of the same mistakes in the future. Shawn, however, is more likely to overcome his misery, go back to functioning as before, and eventually find a new love. Fred will find it more difficult to move on. Research shows he has a higher probability of developing depression. Numerous studies demonstrate that a pessimistic explanatory style (like Fred's) is a risk factor for clinical depression. Depressed individuals are also more likely to view negative events as their fault, permanent, and encompassing all aspects of life.[6] The crucial element that links depressive symptoms with a pessimistic explanatory style is expectation. A pessimistic explanatory style triggers depression by producing negative predictions for the future, which promote negative mood, passiveness, and hopelessness.

Shock, Shock, Shock

The notion of optimistic and pessimistic explanatory styles was put forward by the psychologist Martin Seligman. Seligman developed the concept while trying to explain the results of a study he had conducted a decade earlier. In the mid-1960s, Seligman was a young researcher studying animal learning at the University of Pennsylvania. He wanted to examine whether dogs could learn to avoid an adverse situation if they had prior warning. The idea was simple: First he would teach the dogs that a certain auditory tone would be followed by an electric shock. Then he would give the dogs an option to escape the shocks by jumping over a barrier after the tone was presented. Would they learn to do so?

Before I reveal the results of Seligman's now famous study, let's engage in a short mental exercise. Imagine yourself sitting at the center of an empty room. The walls are bare—no paintings, no plants, not even a window. The chair you are sitting on is the only piece of furniture around. Suddenly, as if out of nowhere, you experience a strong electric shock. The high current runs through your skin, muscles, and hair. A few moments later, another shock is administered, then another and another. I need to get out of here, you say to yourself. You try the door—locked; air shaft—too small; jump on the chair—shock; get off the chair—shock, shock, shock. There's no way out of there, and nothing you do seems to stop the pain. The shocks continue whether you bang on the wall, stand on your head, or lie on the floor. You sit on the chair, feeling terrified and miserable. A couple of hours later, for no apparent reason, the shocks cease and the door cracks open. You sigh in relief and run out.

Your euphoria, however, is short-lived. The next day, to your horror, you find yourself alone in yet another unfamiliar room. It is not the same room—this one has paintings on the walls and a gray carpet on the floor—but sure enough, within minutes dreadful electric shocks start. What do you do?

Think of your response for a few moments while we return to Seligman's dogs. As you recall, Seligman wanted to see if his dogs would be able to escape electric shocks (like the ones you have just imagined) after learning they were predicted by a warning tone. He first put the dogs in harnesses and presented a tone followed by a shock, then another tone followed by a shock, and then another. Soon enough, the dogs would whine whenever they heard the tone, indicating they knew that a shock was coming. While harnessed, the dogs could not do anything to escape the shocks.

Next, Seligman took the dogs out of their harnesses and put them in a box with a low barrier that they could easily leap over. He presented the tone, but to his surprise, the dogs did nothing. They did not attempt to jump out; they simply lay on the floor, howling. Seligman knew that the dogs had learned to anticipate the shock when they heard the tone. Why, then, were they not escaping when the warning sound was presented? There was a clue to what was going on: Only the dogs that had previously been put in a harness acted passively; dogs that had never been put in a harness quickly learned to jump over the barrier to avoid the shocks.[7] It seemed that the dogs that had been in a harness before assumed they once again had no control over adverse outcomes. Even though they were put in a new environment, one from which they could escape the pain, they did not even try.

There were certain aspects to the dogs' behavior that resembled a well-known human condition. The dogs' passiveness, lack of assertiveness and exploration, negative mood, whining, and general helplessness reminded Seligman of patients suffering from depression. The animals also ate less and lost weight, just as depressed patients do. These similarities made Seligman wonder whether clinical depression is caused by a perceived absence of control over outcomes. People with depression, he hypothesized, learn from past experience to behave helplessly. Therefore, even in situations where negative outcomes are avoidable and positive ones attainable, they do not attempt to shape their destiny and thus are less likely to shun harm and achieve desirable results. This, in turn, feeds their depression further. Seligman named his theory *learned helplessness theory,* which ultimately became a dominant model of depression.[8]

Now let's return to that dreadful shock room. Imagine once again that you are there. You are standing alone on the gray carpet when another shock is administered. You feel the electricity traveling through your body. What do you do? Do you wait until the shocks stop? Or do you search for an escape route?

In our hypothetical scenario, most people would certainly attempt to find a way out of the *first* room. However, would they try to escape from the *second*? Or would they assume that room number two was just like room number one and that yet again the door would be shut and the air shaft would be too small to allow escape? Would they learn to become helpless and not even try to avoid the shocks? The answer would vary from person to person. As it turns out, it would also vary from dog to dog.

In Seligman's experiment, not *all* dogs learned to be helpless; not all dogs showed depressive symptoms. A minority of the dogs that had previously received shocks while harnessed went on happily to jump over the barrier to avoid the shocks when given an opportunity. The dogs expressed individual differences, like humans do. Just as some dogs did not learn to be helpless, some humans (like our friend Shawn) will experience severe blows—the death of a loved one, sickness, unemployment, bankruptcy, heartache—and yet put the pieces of their shattered lives back together and move on. If Shawn were a dog in Seligman's experiment, he would probably be the first to leap out of the shock box. He would not assume that a certain situation in which he found himself would apply to all others.

On the other end of the spectrum, we find a small number of dogs that act helplessly even without prior exposure to uncontrollable shocks. As you may recall, Seligman observed that most dogs that had not experienced unavoidable shocks in harness would quickly learn to leap out of the shock box after the first couple of shocks. However, he also found that 5 percent of the dogs would not. They acted helplessly and took the pain for no apparent reason, a passive response that resembled the behavior of individuals who are vulnerable to depression—people like Fred.

Seligman believes that people who think like Fred can learn to think like Shawn; that an optimistic explanatory style can be taught and implemented, even in individuals who are prone to interpret the world in a pessimistic manner.[9] To do so, Fred would first be required to identify the adverse event (that's easy—Sabrina's sudden departure), his interpretation of the event ("I am to blame for Sabrina's leaving because of my unbearable personality"), and the consequences of this interpretation ("I feel miserable and helpless; I am no longer productive at work"). Then Fred would need to consider the evidence in support of, or against, his interpretation ("I have many good friends who love me. I get along with them quite well. That must mean my personality is not that bad") and think of other explanations for why things went wrong ("Sabrina had different goals in life, and we did not communicate well"). Finally, Fred would have to reconsider the implications of the breakup ("Sabrina's leaving does not necessarily mean I will die poor and alone") and consider the benefits of moving on from the failed relationship. Once Fred successfully followed these steps, new hope might arise.

There is evidence that altering a person's cognitive style by using training and talk therapy reduces the likelihood of suffering from depression and enhances physical health. For example, in one study Seligman identified a group of college students with a pessimistic explanatory style. He then conducted training sessions with half of the students, offering them techniques for adapting an optimistic explanatory style; the other students (the control group) were not given training. Months later, students who had received training had fewer self-reported symptoms of physical illness and fewer doctors' visits than students in the control group.[10]

The Right Number of Tennis Balls

Cognitive therapy is one route Fred might choose to take in order to battle his growing depression. In addition, or as an alternative, he might decide to use antidepressant medication (if he was to choose antidepressants, he would not be alone—about 27 million Americans take these drugs).[11] What Fred might not be aware of is that antidepressants would ultimately change the way he thinks—the way he processes and interprets the world around him—in a very similar manner to cognitive therapy (such as Seligman's learned optimism training). How do antidepressants alter a person's perception?

The most commonly prescribed antidepressants are drugs that enhance the function of the neurotransmitter serotonin. A neurotransmitter is a chemical that enables communication between neurons in the brain. It is released into the space between two neurons (known as the synaptic cleft) by one neuron and binds to the receptors of the receiving neuron. You can imagine the neurons as two kids playing tennis, and the neurotransmitter as the ball. One person, William, is serving to the other, Henry. In cases when the ball does not reach Henry's side of the court, William will run to grab the ball back and serve again. The majority of antidepressants, such as Prozac, are selective serotonin reuptake inhibitors (SSRIs). What SSRIs do, in this parallel universe, is inhibit William's desire to go get the balls back (inhibit what is known as reuptake). Instead, William will take another tennis ball from his bag when it is his turn to serve. As a result, more tennis balls will be floating around on the court between William and Henry. Some of the balls will eventually reach Henry, and he will pick them up and put them in his bag.

Back to the brain: As a result of taking SSRIs, the level of serotonin in the synaptic cleft is increased, and there is more serotonin available for binding with the receptors of the postsynaptic neuron (or Henry). The drugs are "selective" because they predominantly affect the function of serotonin rather than other neurotransmitters such as dopamine or noradrenaline. Don't get me wrong: Depression is not caused by the malfunction of a single neurotransmitter. On the contrary, both dopamine and noradrenaline play an important role in depression, and some drugs do indeed target their function, but these are less commonly prescribed than SSRIs (dopamine and its involvement in expectations of reward are discussed at length in chapter 8).

Many people assume that antidepressants directly affect a person's mood—pop one in and you magically become happy. This is not the case. Here is what you will not find in the information leaflet for Prozac: Rather than altering an individual's mood, what antidepressants do is change cognitive biases.[12]

People with a propensity for depression tend to be biased toward negative stimuli.[13] When they walk into a crowded party, they will orient toward individuals with fearful or angry expressions. Later, they will remember negative social interactions (spilling red wine on someone's white dress) better than positive ones (having an interesting conversation with the woman wearing the stained white dress). They will also classify ambiguous interactions as negative (she was not genuinely interested in the conversation; she was just being polite). Such negative biases in processing information result in a negative interpretation of life experiences, leading to low mood and pessimism.

Antidepressants change this pattern; they reinstate positive processing of information.[14] After taking medication, depressed patients start orienting more toward happy faces and other positive stimuli, and they also remember them better. At first, this will not enhance their mood, but after a few weeks of processing more of the good and less of the bad and ugly, the world seems more inviting and mood is enhanced. It takes time for changes in perception, attention, and memory to consolidate and alter a person's emotional state. This is partially why antidepressants do not have immediate mood-enhancing effects; it takes a few weeks before depressive symptoms are visibly reduced.

Given that most antidepressants work by altering the levels of serotonin in the brain, it may come as no surprise that a paper published in the prestigious journal *Science* in 2007 revealed that a gene coding for serotonin function predicted a person's likelihood of suffering depression.[15] The gene identified is one that encodes for the serotonin transporter, which removes serotonin from the synaptic cleft (or tennis balls from the court). The serotonin-transporter gene has alleles that can come in either a long version or a short version (an allele is a DNA sequence in a particular gene). Each gene has two alleles; whether someone has two long versions of the serotonin-transporter gene allele, two short versions, or one of each will determine the expression and function of the serotonin transporter, and therefore the function of serotonin itself. In individuals with the short version of the allele, the serotonin transporter functions less efficiently. Such individuals are also twice as likely to suffer from depression—but only if they have experienced a stressful life event such as unemployment, divorce, bankruptcy, or health problems.[16] In other words, low efficiency of the serotonin transporter does not directly enhance a person's susceptibility to depression. Rather, it makes a person less resistant to stressors, and therefore makes it more difficult to overcome life's downturns (much like having a weak immune system).

Humans are not the only species that display depressive symptoms in response to adverse events. We saw that earlier: Seligman's dogs acted depressed after receiving shocks that were out of their control. The association between genes that code for serotonin function and depressive behavior is not uniquely human, either. Although Seligman never took saliva from his dogs to test for their genetic makeup, studies of other nonhuman animals suggest that the relationship between serotonin function and the blues travels way down the evolutionary ladder—as far back as mice.

You may think of mice as being very different from our own species. What could a mouse possibly tell us about a condition as complex as depression, a condition most of us consider reflects the essence of human fragility? There are many distinctions between mice and men: Mice are smaller, they possess a long tail and pointy ears, and they are often eaten live by large birds—something that rarely happens to humans. It is difficult to imagine a mouse contemplating the meaning of life or agonizing over lost love (like humans, however, these mammals are frequently found in the kitchen in the middle of night, searching for leftovers). Regardless of our dissimilarities, mice are the laboratory animals most closely related to humans to which genetic engineering can be applied with relative ease. Genetically engineered mice are referred to as "knockout mice"—they have had a gene "turned off." By inactivating a specific gene in a group of mice, scientists can examine how the behaviors of these mice differ from those of mice that have not been genetically engineered. In this way, the particular processes that are mediated by that gene can be identified.

To examine the role of the serotonin-transporter gene, the gene was disrupted in a group of mice. At first, the knockout mice did not seem any different from the control mice (those that had not been genetically engineered). However, after putting the mice in a stressful environment, differences started to emerge. The mice with the disrupted gene showed an enhanced response to stressors. This was observed both behaviorally and physiologically; they expressed more fearful behavior and their stress-hormone levels were higher.[17]

The reason for studying genetically engineered mice is not to learn about these small creatures; rather, scientists fiddle with the genetics of mice in hopes of learning something about humans. Will humans with less efficient serotonin function also show enhanced physiological response in stressful situations? The answer is yes. Results of a brain-imaging study revealed that people with the short version of the allele had higher activation in the amygdala in response to fearful and angry faces, as well as in response to negative words (such as *cancer*) and disturbing pictures (such as those of mutilated bodies).[18] Why does the amygdala of people with short alleles overreact in response to stressful events?

The amygdala is a structure deep in the brain that processes emotional stimuli. It is also involved in generating physiological responses to these stimuli. Amygdala activity is regulated by parts of the frontal cortex, in particular the anterior cingulate cortex (ACC). In individuals with a short allele of the serotonin-transporter gene, there is reduced connectivity between the ACC and the amygdala.[19] This means that the two structures are not as good at communicating with each other. As a result, the ACC is less efficient in reducing fear and stress responses in the amygdala. This is particularly problematic when fear responses, which may have been called for previously, are no longer appropriate. For example, if I put you back in the shock room, you would probably react with fear and anxiety because you would be expecting electric shocks at any moment. Your pulse would rise, sweat would start dripping from your forehead, and your mind would be hijacked by one thought only: When will the shock arrive? If ten minutes elapsed and no shocks were delivered, you would start calming down. An hour later, you would be completely relaxed, humming away and thinking about what you might have for dinner. This is known as *fear extinction*—the process of learning that something that was previously threatening no longer is. Fear extinction involves the regulation of amygdala activity by the ACC. As the connectivity between these structures is relatively impaired in carriers of the short allele, these individuals will be less capable of extinguishing their fear. They will be more likely, therefore, to maintain high levels of anxiety and be prone to depression and other mood disorders.

As depression is not an illness involving malfunction of one neurotransmitter alone, it is also not a disease related to deficits in only one or two brain structures. It reflects, like many other mental disorders, a system failure. This system includes brain regions I discuss in this book, including the hippocampus (which has an important role in memory) and the striatum (which is involved in motor function, reward processing, and generating expectations of pleasure and pain), as well as other brain regions I focus on to a lesser degree, such as the thalamus and the habenula. Abnormal activation in these regions, and disrupted communication between them, is often observed in depressed patients. It has been discovered, however, that to cure depression, it is sometimes sufficient to target only one region.[20] Changes in a single brain area can modify the function of other connected structures.

The First Day of Spring

How can doctors alter the function of a target brain region? The answer is deep brain stimulation (I briefly mention this technique in chapter 8). The technique is invasive; it involves implanting electrodes into the brain of the patient and then stimulating the brain tissue at high frequency. The electrodes are connected to a small battery pack, which is normally implanted close to a patient's collarbone. The system is controlled by an external device, so it can be turned on or off to deliver electric stimulation.

Deep brain stimulation is a well-known treatment for Parkinson's disease. For treating depression, however, it has been attempted on only a relatively small number of patients. Helen Mayberg, now at Emory University, pioneered the treatment together with her colleagues while working at the University of Toronto. She was attempting to treat severely depressed patients desperate for a cure. These patients had already tried a variety of treatments (psychotherapy, antidepressants, electroconvulsive therapy) without success. Mayberg's idea was to target the very brain region that most consistently showed irregular function in depressed patients—the subgenual cingluate cortex, which is part of the ACC.

As Mayberg herself testifies, she did not know what to expect.[21] This area of the brain had never before been operated on in such patients. Thankfully, what she was about to observe was better than anything she'd dared hope for. Her first patient, a women suffering from years of severe depression, was on the operating table. Her head was secured in a metal frame to keep it perfectly still while the neurosurgeon drilled two narrow holes through her brain—one on each side. The patient was wide awake; she was aware of the fact that tiny foreign objects were about to be inserted into her brain, but she could not feel a thing. During neurosurgery, it is important to have the patients stay awake so that their cognitive and motor functions can be monitored. The doctors had to make sure that the patient's speech was normal, that she could recognize their faces as well as move her fingers and toes. They also wanted to know what the patient was thinking and how she was feeling.

The surgeon inserted an electrode through each hole to reach the white-matter tracts of the subgenual cingulate. He then administered a small current. This was the critical moment. Would stimulating the white-matter tracts alter the patient's cognition? Would it change her mood? The team in the operating room turned silent in anticipation, and then . . . nothing. The patient did not notice a thing. Yet not all was lost. Each electrode had a few contacts with the brain, and although stimulating the first contact did not result in any noticeable changes, the second contact was the jackpot. The surgeon administered a moderate amount of current through the second contact. "Did you just do something?" asked the patient. "I have this intense sense of calm or relief. It's just so hard to describe; it's like that first day of spring when the crocuses just pop up through the snow." That caught Mayberg's attention. "Wait a minute. Are you seeing crocuses?" she asked. "No, no, I'm trying to think of something that evokes this kind of emotional state, this kind of calm and satisfaction," the patient responded. "It was as if she realized that first day when you walk outside in spring and see the flowers, it's a sense of renewal, it's the start of spring. It was the most amazing and poetic statement," Mayberg recalled years later.[22] Not all patients were as poetic as this one, but Mayberg could observe a change of emotional state in most. She could view this transformation in their expression; when the second contact was turned on, their faces would suddenly relax. One patient described it as an ability to turn attention away from the misery inside toward the people and events on the outside.

Mayberg emphasizes that what her patients experienced was not a sensation of becoming happy. Rather, it was one of regaining control. Before the operation, the emotional control system was, as Mayberg puts it, "hijacked." After deep brain stimulation, it was released, and there was a state of calm and relief. "There is something about this small amount of current, in this very specific place, that allows the system to just reequilibrate," she says.[23] In Mayberg's patients, once an efficient interaction between the monitoring frontal cortex and the subcortical emotional system was restored, their depression was cured. Mayberg managed to help two-thirds of her patients, and the benefits were long-term.

It is interesting that when examining optimism, my colleagues and I found that activity in the area targeted by Mayberg predicted the degree of optimism in our subjects.[24] You may recall that in that first brain-imaging study on optimism, we saw enhanced connectivity between the ACC and the amygdala in healthy, optimistic individuals when they imagined positive future events (such as a sunny ferry ride), compared to negative events (such as losing a wallet).[25] So while depression is associated with impaired connectivity between the ACC and the amygdala (and thus reduced ability for regulating emotions) as well as enhanced attention to upsetting stimuli, optimism, on the other hand, is associated with enhanced connectivity between the ACC and the amygdala (leading to efficient emotion regulation) and greater attention to positive stimuli. As you will remember, depression is also related to short alleles of the serotonin transporter. Is optimism related to long alleles of the serotonin-transporter gene?

Indeed, studies show that individuals with two long versions of the allele tend to be more optimistic. They score higher on the optimism trait scale,[26] report higher levels of satisfaction with life,[27] and, according to a study conducted at Essex University by the psychologist Elaine Fox, show a tendency to attend to the bright side. In her study, Fox presented participants with photos of either positive stimuli (such as someone smiling or ice cream) or negative stimuli (such as people frowning or an insect) alongside neutral photos. She found that the attention of people with two long alleles was captured more often by positive stimuli than by neutral stimuli, and less often by negative stimuli. Fox concluded that this bias effectively protects people from clearly perceiving the negative aspects of life and enhances the tendency to perceive the rosy side. This positive bias was not detected in participants with one or two short alleles.[28]

Our tendency for misery or ecstasy is partially determined by our genes, yes, but it is also related to our environment, to our health, and to our unique experiences (as well as to many other factors). It is the combination of all of these that will ultimately trigger depression, or protect us from it.

CHAPTER 7

Why Is Friday Better Than Sunday?

The Value of Anticipation and the Cost of Dread

119.5 seconds. That is the time it takes to pour and serve a perfect pint of Guinness.[1] The glass is held at exactly a forty-five-degree angle under the tap. Next, the famous double pour: The glass is filled up three-quarters of the way and left to settle. Once the bubbles settle, creating a creamy head, the glass is topped off.[2] The double-pour ritual originated back when the dark Irish liquid was served straight from the barrel. The bartender would prepare glasses filled three-quarters of the way with older stout, then leave them to stand. When a customer came in to order a pint, the bartender would top up the glass with fresher, younger stout, which produced the foamy top.[3] Nowadays, a pint of Guinness is no longer a mix of old and new stout, yet the double-pour tradition remains. In fact, it is mandatory. Guinness has instituted the "perfect pint training program," which ensures that wherever in the world Guinness is served, the double-pour technique is used, creating the perfect creamy head, which is about one-third to one-half inch in depth.[4] Did the double-pour technique survive through the years simply because it creates a foamy head that does not overfill the glass? Not at all. The careful 119.5-second pour generates something much more important. It produces what some consider the most vital aspect of the Guinness experience—anticipation.

In November 1994, dramatizing the thrill experienced while one eagerly waits for the cold stuff to settle, Guinness came out with one of its most successful advertisements. "Anticipation," as the ad was called, featured a customer dancing around in excitement while a bartender poured him a pint. The ad ran for only sixty seconds. Nevertheless, those sixty seconds, and the campaign that followed—the famous "Good things come to those who wait"—had a huge impact. Guinness's sales went through the roof and brand recognition skyrocketed.

The Value of Anticipation

What the marketing people at Guinness tapped into was a central aspect of human nature we tend to overlook—the joy of savoring. Sometimes *expecting* a good thing is more pleasurable than actually *experiencing* it. Consider the hours blissfully spent daydreaming about an upcoming vacation—we get our money's worth before we even board the plane. Or consider the excitement while getting ready for a keenly awaited date, all sorts of future scenarios running through our mind; the childhood enthusiasm weeks before Halloween or an upcoming birthday; the exhilaration before reuniting with a loved one. I could go on and on.

Although we have all felt the joy of expectation, we rarely consider its value explicitly when making decisions. How many of us have ever said, "Well, if you take into account the weeks of pleasure I will obtain from anticipating the trip to Venice, I guess a thousand dollars is not too much to spend on a weekend abroad." While we may not believe anticipation is a source of satisfaction in and of itself, our actions reveal otherwise. Contemplate this scenario: Your loving spouse has decided to buy you tickets to see your favorite band in concert as a birthday gift.

The band will be in town for a few weeks. "When would you like to go?" asks your significant other. "Tickets are available for tonight, tomorrow night, for a show in two days' time, five days' time, or next week." Which date would you pick?

When given the choice, people would rather wait a bit for a good thing than have it immediately. Most of us would choose to go to the concert later in the week rather than right away. In a survey conducted by George Loewenstein, an economist at Carnegie Mellon University, undergraduate students were asked how much they would pay to receive a kiss from a celebrity of their choice.[5] Imagine a passionate kiss from X (this is where you fill in the blank—Angelina? Brad? Patrick Dempsey? Uma Thurman?). After you make this tricky decision, write down how much you would be willing to pay to receive a kiss from that person immediately, in one hour, three hours, twenty-four hours, three days, a year, or ten years.

Loewenstein found that on average people would pay more to receive a kiss from a celebrity in a year than to receive it immediately. An immediate kiss would leave zero time for anticipation. We would be giving up the thrill of the wait, the pleasure derived from imagining the expected kiss, considering how and where it would take place. However, if the kiss was expected in a week's time, one could frequently muse on the upcoming event. With each reflection, a brief moment of joy would be produced. Students were even willing to pay a bit more to receive the kiss in a year's time rather than in three hours. They were not, however, willing to wait ten years. Who knows if our object of desire will be as desirable in a decade? The most preferred waiting time was three days, reflecting a balance between the pleasure of anticipation and impulsivity (we will return to the role of impulsivity later in the chapter).

The fact that people decide to wait for a rewarding event rather than receive it immediately suggests that we derive pleasure from contemplating something that might happen later. Even if our current state is negative (for example, we are working late at the office on a Friday evening), we can feel happiness simply by thinking of the weekend ahead. In fact, when you ask people to rank the days of the week in order of preference, Friday is ranked higher than Sunday, although Friday is a workday and Sunday is not.[6] Would people rather work than play? Not quite. Saturday, which is also a day of play, is ranked above both Friday and Sunday.

So why do people prefer Friday to Sunday? The reason is that Friday brings promise—the promise of the weekend ahead and all the activities (or non-activities) we have planned. Sunday, while a day of rest, does not bring with it the joy of anticipation. To the contrary, although we may be having a picnic in the park or strolling around town, these delightful activities are marred by the anticipation of the full workweek ahead. Whether good or bad, our emotional state is determined both by feelings that are triggered by the world at present and those generated by our expectations of the future.

The Cost of Dread

Consider another scenario: You are at the dentist's office for your annual checkup. While examining your teeth, your dentist concludes that, unfortunately, you need a root canal. There is no patient scheduled after you, so the dentist can conduct the procedure right away. Alternatively, you can be penciled in for later that afternoon or for next week. What do you do? When it comes to adverse events, most of us choose to get it over with as soon as possible. The reason is simple: We want to avoid the dread that comes with anticipating pain. Instead of spending our time worrying and fearing, we would rather face the pain immediately and be done with it.

Indeed, when participants in Loewenstein's survey were asked how much they would pay to avoid a 120-volt electric shock that could be delivered immediately, in three hours, twenty-four hours, three days, a year, or ten years, the students indicated they were willing to pay the most to avoid a shock in ten years. In fact, they were willing to pay almost double to avoid the electric shock in ten years than to avoid an immediate shock.[7] In another study, where shocks were actually delivered, some participants were so keen to avoid the dread, they opted to receive a larger electric shock immediately rather than a smaller, less painful one later.[8]

This decision may seem irrational. Traditional economic theorists would certainly claim so. According to these classic decision-making models, humans are rational agents who try to optimize expected utility.[9] *Utility* is an economic term that refers to the relative desirability of an object, or to the satisfaction we can derive from it. Electric shocks are neither desirable nor satisfying. So if on a scale of 1 (not at all painful) to 100 (so painful, one would rather die) we expect a 120-volt shock to be about 40, and are willing to pay one hundred dollars to avoid it, then we should be willing to pay one hundred dollars to avoid it now and one hundred dollars to avoid it in ten years. This is because in both cases we expect the shock to be painful to the same degree. Now, let's say you are willing to pay one hundred dollars to avoid the shock today and two hundred dollars to avoid it in a year (although in both cases the expected pain is 40 on the pain scale). This is a violation of rational behavior. Or is it?

What classic decision-making theories fail to account for is the negative value of dread (many modern economic theories still fail to do so). If we take anticipation into account, the behavior described above seems perfectly rational. Yes, the expected pain from a 120-volt shock is about 40 on the pain scale whether we receive it now or in a decade, and avoiding it is worth about one hundred dollars either way. However, one has to consider the distress that would be caused by anticipating the shock over ten years. Avoiding ten years of dread may be worth an extra one hundred dollars. It is therefore completely rational to be willing to pay double to avoid an adverse event in the future. Receiving the shock now means no time to contemplate possible negative consequences, none of the sinking-gut feeling that would emerge every time the thought of the impending shock came to mind.

Not only is it sensible to pay more to avoid a negative event in the future than to avoid one in the present; it is foolish not to. The negative influence of anticipating an unwanted event on our physical and mental health can sometimes be worse than the effect of experiencing the event. This phenomenon was observed in the mid-1970s in employees of two manufacturing plants in the United States. One plant was located in a large metropolitan area, the other in a rural community with a population of three thousand. The first was a paint-manufacturing plant; the second manufactured display fixtures used by wholesale and retail concerns. The employees in the two plants were machine operators, laboratory assistants, clerks in shipping departments, assembly-line workers, and tool-and-die workers. On average, the employees had worked at their respective plants for twenty years. Sadly, both plants were scheduled to shut down, and all workers were about to lose their jobs.

For months, the men came to work knowing that in only a few weeks they would be unemployed. Anticipating the loss of their workplace, where they had spent most of their days for the past two decades, was stressful. The anxiety was triggered to a large extent by the uncertainty of what lay ahead. How would they cope with unemployment? Would they find another job?

Scientists who followed these workers before and after the plants were closed found that the employees experienced more days of illness before the plants were shut than during the weeks of unemployment that followed.[10] The anxiety induced by anticipating the loss of their jobs damaged their health and well-being. Ironically, once unemployed, the workers became healthier. The uncertainty of how life would be without a job was lifted. Anxiety was reduced, and attention turned to finding a new job, rather than worrying aimlessly about what might be.

Anticipating the Fall Alters the Impact

A few years ago, I received a surprise birthday present from a friend—a tandem skydive. To be clear, I had never expressed any interest in jumping out of a plane in midair. Falling through the sky fifteen thousand feet from the ground, reaching speeds of approximately 120 miles per hour, was never on my wish list. All the same, that was exactly what I was about to do.

My friend initially intended for the skydive to be a surprise. I was to be driven to a ranch in upstate New York, where the skydiving school was located, and there my birthday gift would be revealed. After careful consideration, however, my friend decided it might be wise to let me get used to the idea of jumping from a plane in advance, so that I could prepare emotionally. Hence, the surprise was unveiled three days beforehand. I now had seventy-two hours to ponder the upcoming jump. Depending on your personal preferences, those seventy-two hours could be viewed as a time of pleasurable excitement or as days of pure dread—the latter was my experience. I was walking around town with a death sentence hanging over my head. I turned to the Internet for help.

Entering the words *skydiving* and *perish* on Google revealed that approximately thirty people die every year in the United States while skydiving. This may seem a large number at first; however, considering the 2.5 million jumps that take place in the United States every year, it is, in fact, a small percentage. A closer investigation revealed that death or serious injury from tandem jumps (jumps in which you are tied to an instructor, as I would be) are particularly rare. This was encouraging. Although I had experienced three days of stressful anticipation, having that time allowed me to acquire knowledge regarding the threatening event. My fear was reduced by the information I had gathered, allowing me to enjoy the experience more (yes, I admit I ended up taking pleasure in the rush of adrenaline).

Now, imagine all this had taken place pre-Google. In fact, imagine it had been impossible for me to gather any information on skydiving—that I had no access to anyone who had jumped before, or anyone who knew anything about the experience at all, and no statistics at hand. I would probably have spent three anxious days picturing the worst. Obviously, anticipating a potentially adverse event is unpleasant, but does it also affect the way we experience the event itself? Does the dread of a root canal make it more adverse? Does the fear of a shock make it more painful?

A neuroimaging study published in the journal *Science* in 2006 suggests this is the case. The scientists reported that people who dreaded an upcoming shock found the actual shock worse when they had to wait longer for it.[11] In other words, if you are awfully anxious about getting that root canal, better do it as soon as possible. Not only will you avoid unpleasant anticipation, but the whole experience may seem less painful now than it would in a week. Interestingly, the amount of dread experienced before the shock did not alter brain activity during the shock itself. It was only before the shock that increased dread was related to heightened activity in the brain's "pain matrix."

The pain matrix is a network of brain regions that are associated with processing different aspects of the pain experience. This network includes the somatosensory cortex, which responds to the physical aspects of pain, as well as areas thought to respond to emotional processing, such as the amygdala and rostral anterior cingulate.

During the time of anticipation, people who dreaded the shock a lot had greater activity in brain regions that usually process the physical intensity of pain (such as the somatosensory cortex). This means that the brain responded similarly to the anticipation of the shock and to the shock itself. Anticipation seemed to mimic the actual experience of pain. Dreaders also had greater activity in areas thought to modulate attention to pain, which suggests that dread enhanced attention to the physical aspects of the expected pain. If anticipating an adverse event activates areas of the brain that normally process the physical experience of pain, it is hardly surprising that anticipating a painful event has a negative effect on our well-being similar to that of actually experiencing it.

In a like manner, anticipation of a pleasurable event seems to activate neural systems that are also engaged while actually experiencing the enjoyable event. For example, a study my colleagues and I conducted showed that when people imagine a future vacation, the striatum—a brain region that also responds to actual rewards, such as food, sex, and money—is activated.[12] This, of course, does not mean that imagining a vacation or a juicy cheeseburger is the same as actually being there on the beach or eating a burger. Yet sometimes the pleasure we obtain from thinking about a burger comes close to the pleasure we feel while sinking our teeth into one.

Not all burgers are created equal. We may enjoy a moist burger with lettuce and cheese more than a dry one without toppings. We are also more likely to enjoy a burger when we are hungry than when we are full. But do these differences matter when we are only anticipating the burger? What determines how pleasurable we find anticipation?

There are a few critical factors.[13] First, the tastier the burger is expected to be, the greater the pleasure from anticipating it. If we are about to have green salad and peas, we will probably not derive much joy from anticipating our meal (unless, of course, we find green salad and peas pleasing). Second, the more vividly we are able to imagine an event, the greater the pleasure from anticipating it. If we are unable to construct a detailed image of the burger, one that includes the smell and texture of it, anticipating it will be less enjoyable. Third, how probable you think the event is influences how joyful anticipating it will be. If we think there is no chance we will be able to leave the office for lunch, we will not derive much joy from imagining the unattainable burger. Finally, time matters. As lunchtime approaches, the excitement about the upcoming meal becomes greater and greater.

The same principles apply when anticipating adverse events. The dread of anticipating a root canal will be determined by how painful we think the procedure will be, how vividly we can imagine the noise and vibration of the drill, how likely it is that we will need a root canal, and when we expect it to take place.*

Let's think back to the dancing Guinness patron awaiting his drink. Does this particular customer perceive the Guinness pint to be half full or half empty? If the patron is a "glass half full" type of person, will he derive more pleasure from anticipation?

The optimism bias is, by definition, our tendency to overestimate the probability of positive events and underestimate the

*Notice that the nearer in time we believe we are to when the root canal will take place, the greater the dread. People choose, however, to get the root canal sooner rather than later because if they sum up all the moments of dread from the present time to the time of the procedure, the total amount of experienced dread is greater for a later root canal than for one that takes place sooner.

probability of negative ones. That's not all. Optimistic people imagine positive future events with greater vividness and detail than negative events and imagine them occurring nearer in time. So the more optimistic you are, the more likely you are to imagine positive events as nearer in time, with greater detail, and with higher probability than negative events.[14] Optimism thus modulates the same factors that influence the value of anticipation: predicted pleasure, vividness, the expected time of the event, and its probability.

For desirable events, such as a foamy pint of Guinness, optimism increases the joy of anticipation by (a) increasing our expectation that we will receive our pint sooner rather than later, (b) enhancing our ability to imagine the cool feel of the glass and the smooth flow of the substance, and (c) increasing the perceived likelihood of receiving the stout. Pessimists, on the other hand, may have trouble vividly imagining the black stuff, will predict that the bartender will take a long time to deliver the stout, and may worry that he will run out of Guinness altogether. Although our pessimistic customer may enjoy the pint when it is finally delivered, he will be deprived of the 119.5 seconds of pleasure that precede consumption, and it is unlikely he will dance around the pub while waiting for his drink.

The same rationale holds for negative events such as unemployment. An optimist will estimate the likelihood of getting fired as low, will have a hard time imagining the scenario in detail, and will predict that if the event does happen, it will be far, far in the future. The result? Dread, anxiety, and stress are reduced. A pessimist, on the other hand, will be sure she'll be the next employee to go, possibly tomorrow, and will imagine the whole thing in gruesome detail. Anticipating unemployment (which may never come) is not only aversive; it will induce stress that will negatively affect physical and mental health.

It seems plausible, then, that the optimism bias developed

partly because optimism maximizes the pleasure we obtain from anticipating a good thing and minimizes the adversity of anticipating a bad thing. If a burger is equal to one hundred "units of joy," an optimist will obtain more units of joy just from *anticipating* the burger than will a pessimist, ultimately increasing the pleasure derived from the burger and enhancing well-being. So, if optimists find anticipation more pleasurable than pessimists do, are they more likely to delay gratification in order to increase the length of anticipation?

The "Survivor" Dilemma

The answer is complex. This is because the value of anticipation is not the only factor that determines when we decide to indulge. There is at least one other crucial factor: *temporal discounting.* Temporal discounting is the tendency to value the present more than the future. For example, if you are given a choice between receiving $100 today or $100 in a month, you will most likely take the money today. That's an easy decision. However, given a choice between $100 today and $105 in a month, what will you do? Most people would rather have $100 today than $105 in a month. Some people would even take $100 today over $150 next month.[15]

At first glance, and indeed at second, it seems that temporal discounting steers our decisions in a direction opposite from anticipation; it drives us to consume goods as soon as possible and delay pain until sometime in the unforeseeable future. This is not only because we tend to value the here and now more than the there and later but also because we (correctly) perceive the future as uncertain.[16] We would rather eat our chocolate cake now than save it for later, because tomorrow we may find that the cat has already gobbled it up. We may decide to delay cleaning the house until next week, because by then our spouse may have already gotten around to it. However, if we know for certain that our spouse has no intention whatsoever of cleaning the house, we may decide to get to it as soon as possible and be done with it. Likewise, if we had a crystal ball that revealed that our scrumptious chocolate cake would stay fresh in the fridge and would not be eaten by anyone else, we might delay gratification a bit longer to prolong anticipation. However, even if we lived in a magical universe with a fully predictable future, people would still discount the future to some extent. If we did not discount the future at all, we would never get around to actually eating the cake or opening that precious bottle of wine we had been saving in the cellar. We would keep delaying gratification again and again in order to extend pleasurable anticipation.

Anticipation and temporal discounting pull us in different directions until an equilibrium is reached. While the pleasure of anticipation makes us patient creatures, temporal discounting makes us impulsive. What we will ultimately decide to do reflects the balance between these two factors. When the value obtained from anticipating a future reward is greater than the value of consuming the reward at present, we will delay our indulgence. However, when our desire to have the gourmet Swiss chocolate is greater than the joy of savoring it, we will tear into the wrapper.

Many factors play a part in whether or not we will ultimately savor a product. Things that are available infrequently (such as an expensive bottle of champagne or an annual vacation) may be worth savoring. Something that can be had again and again (such as a kiss from our partner) is consumed whenever we feel the urge. Another factor that determines our decision is whether the object has a greater value now than we expect it to have in the future.

Consider the reality show *Survivor*. In *Survivor*, innocent citizens from developed countries are taken to a deserted tropical island and left to fend for themselves. They are allowed to take only the clothes on their backs; no BlackBerrys, iPods, or can openers are permitted. Not even toilet paper. Contestants can avail themselves anything they find on the island, such as bamboo and edible fruit. They also receive a limited amount of water and food—usually some rice. At the end of their first day, the survivors are famished. They know, however, that while their hunger will only increase with each passing day, their food supply may diminish. Should they save their limited amount of rice for later in the week? Should they divide it into small portions and have a bit every day? Or should they eat it all now and hope for the best?

The difficulty with making such a decision is that it relies upon an ability to predict what, and how much, we will need in the future. This is not an easy problem. The survivors have to consider whether they expect to find additional food on the island and whether they can survive on a few grams of rice a day.

Elaine, an optimistic survivor, reckons that the group is sure to find plenty of berries and coconuts once the sun comes out. They may even be able to catch some fish. She therefore suggests having the rice tonight. That way, Elaine says, the group will have the necessary energy to go out hunting for food tomorrow. Patrick is less optimistic. He does not believe the group members will be successful in finding nourishment in the near future. He argues they should preserve the food as long as physically possible.

Elaine, who favors eating now rather than saving for later, may appear on the surface to have a high "discounting rate"—this is economic jargon for saying she considers the present to be more important than the future. Economists assume that people with high discounting rates are impulsive. These individuals are thought not to be concerned about the future as much as they should be. They don't have savings accounts, and they might indulge in unhealthy practices such as drinking and smoking, which carry penalties in the future. Elaine is not necessarily impulsive or unmindful of the future. Her preference to eat now rather than later is based on her optimistic expectations of what is yet to come. Under those rosy assumptions, it may be reasonable to consume the rice now rather than save it for tomorrow.

To formally test whether optimism alters discounting rates, scientists at the University of Amsterdam conducted a study in which participants were asked to imagine being happily employed by a large company.[17] The company was doing especially well and the executives decided to reward all employees with a raise. The employees had a choice: They could either receive a raise that would begin immediately and last for twelve months, after which their salary would go back to its current rate, or they could receive a raise that would last for thirty-six months but would not go into effect for another twelve months. What should they do?

Many participants selected the short-term raise, which would go into effect immediately, rather than the long-term raise, which wouldn't go into effect for a year. Why would they do that? Why would they choose a twelve-month raise over a thirty-six-month raise when the latter was obviously worth more? The Dutch researchers hypothesized that people picked the short-term immediate raise because they held optimistic opinions about what would follow. For example, the participants may have believed that the company would continue to do well and would eventually decide to give the employees extra bonuses at the end of the twelve months. They may also have considered that the immediate raise could be invested, making them more money in the long run.

To test this hypothesis, the experimenters conducted the study again, this time with a new group of participants. Only they added a small twist. The participants were presented with a scenario that was much more constrained than the one given to the first group. They were told the company would give the raise only once, discouraging the idea that gains would be followed by additional ones. They were also informed that the raise would be adjusted for interest rates and inflation. The participants therefore had less room to generate alternative future scenarios of their own. Under these circumstances, in which the future was not open to rosy interpretation, more participants chose to wait for the larger bonus than to accept the smaller one immediately.

Another group of participants was told that the company was not doing well and that all salaries would be cut by 10 percent. The employees could select whether to have their salary reduced for the next twelve months or to have it reduced in a year's time for thirty-six months. When the future was relatively unconstrained, participants tended to pick the larger delayed loss rather than the small immediate one. Presumably, participants imagined the company might do better over time and would eventually decide that salary reductions were not needed after all. Participants may also have imagined they could find better-paying jobs in a year, and thus avoid the salary cut altogether. When the future was presented as certain (i.e., when the scenario clearly stated that all cuts were fully guaranteed and that employees could not change jobs), more participants opted for the immediate salary cut rather than the delayed one. The conclusion drawn from this data was that temporal discounting is partially due to people's belief that gains will be followed by more and more gains and that losses will somehow be avoidable in the future. This seems like a reasonable explanation for why we would choose to have our rewards now and delay losses for a future time. There are, however, compelling alternatives.

Even the King of Pop Will Age

One such explanation is related to the way we perceive our future selves. Imagine yourself a year from today. You get up in the morning (Where do you live? Are you renting an apartment? Do you own a house?), get dressed (How do you look? Has your hair changed? What will you wear?), have breakfast (Coffee? Cereal? Toast and jam? Scrambled eggs? Skip breakfast altogether?). You kiss loved ones (children, husband, wife) good-bye and head to work (Do you get there by car? Do you take a subway? Do you work from home?). You arrive at your job (What do you do? Do you work in an office? Are you self-employed? Have you already retired?); then, around 1:00 p.m., you take a lunch break (Are you eating healthy foods? Do you have dietary restrictions due to a physical condition?). At the end of the day, you head out (Are you going home for dinner? Are you spending a night on the town? Will you watch a film? Will you go shopping?).

After you have carefully played out the scenario in your mind, do it once more, only this time imagine yourself in ten years. Take your time. When you are done going through a typical day in your life a decade from now, do it again (this is the last time, I promise), but now imagine yourself thirty years from today. Try to imagine your day with as much detail as you can.

How similar are your future selves to your current self? Is the potential you a year from today more like yourself in the present than the potential you a decade from now? What about the future you thirty years from today? Can you recognize yourself in the person you imagine yourself to be in thirty years? It is not surprising that people feel more connected to their near-future selves than their far-future selves. We do not tend to picture ourselves very differently in a year's time. In a decade or two, though, we certainly expect to have transformed quite a bit. We may consider our future selves in thirty or fifty years to be so different from our current selves that we think of them as different people altogether. If we perceive our future selves to be distinct agents from our current selves—strangers, even—it is not shocking that we choose to indulge in the present and relegate pain to the faraway future. Yes, smoking and drinking today mean increasing your risk of having a range of medical conditions in the future, but the person who will have to endure those illnesses does not seem to be you, and so you are less concerned with this future person than you are with the one reading this book. We would rather have a raise today than in a couple of years, because today *we* will receive the money, while in a few years a fuzzy version of ourselves will be enjoying it. You might be as keen to give your future self the dough as you are to hand it out to a stranger on the street (well, not quite, but you get the logic).

An fMRI study published in 2006 aimed to link the way we perceive our future selves with the way we discount time.[18] While their brains were scanned, participants were requested to evaluate the characteristics of both their current selves and their potential selves ten years from now. Later, outside the scanner, participants were asked to make decisions that would reveal their temporal preferences. Choices such as "Would you rather get ten dollars today or twelve dollars in a week?" and "Would you prefer ten dollars today or fifteen dollars in two months?" were presented. The findings showed that people who tended to choose a small gain immediately rather than a larger one in the future had greater differences in their brain activity when thinking of their current selves versus their future selves. Participants who did not discount the future much had only minor differences in activation when evaluating their current and future selves.

In the spring of 2009, the pop star Michael Jackson, then fifty years old, announced his intention to go on tour for the very last time. The tour was scheduled to include fifty gigs at London's O_2 arena—a show for each year of Jackson's life, which ended unexpectedly before he had a chance to go onstage. Most of us would have thought that Jackson would be financially set for retirement. Surely one of the most successful musicians of all time, who had been performing since the age of five, could retire comfortably, with no financial concerns. This was not the case.

The "This Is It!" tour (as it was called) was supposed to get Jackson out of the deep financial fiasco he found himself in before his death. What drove the King of Pop to leave behind a mountain of debt? Like many of us, his main fault seemed to be overspending and undersaving. "Millions of dollars annually were spent on plane charters, purchases of antiques and paintings," said Alvin Malnik, one of Jackson's advisers. "There was no planning in terms of allocations of how much he should spend. For Michael, it was whatever he wanted at the time he wanted."[19]

Michael Jackson was not alone. In 2005, the savings rate in the United States was negative for the first time since the 1930s.[20] This meant that Americans were spending more than they were taking in after taxes. While Jackson's expenses included about eight million dollars annually for luxuries,[21] his fellow citizens were making more modest purchases—a new car, perhaps.[22] Nevertheless, those acquisitions put their retirement funds at risk.

What drove Michael Jackson and his fellow Americans to a negative savings rate? It may have been a bit of overoptimism. From 2002 to 2006, the value of people's homes was rising sharply. Believing this trend would continue, people felt they could spend more.[23] They were wrong. In 2008, real estate prices fell sharply. Although Americans started saving again soon after, savings rates were not enough to provide a quality of life after retirement that was equal to the one people were accustomed to during their working years.[24]

The second problem was that Michael, like many others, had a difficult time imagining himself at the age of seventy.[25] Even if we manage, with some effort, to imagine our aging selves, most people find the image aversive. Alas, thinking about retirement funds involves thinking about aging. As we would rather avoid the thought altogether, many of us shun planning our financial future.

CHAPTER 8

Why Do Things Seem Better After We Choose Them?

The Mind's Journey from Expectation to Choice and Back

My friend Tim works for an online travel agency. Each Christmas, his company grants him a special bonus—an all-expenses-paid vacation. He can go wherever his little heart desires—Australia, Thailand, Italy, Egypt, Hawaii, Las Vegas—the world is his oyster. And so, every year around Christmas, Tim agonizes over the same problem: Where should he spend his winter vacation? "Panama," he declares one day; "the weather is gorgeous there this time of year." Twenty-four hours later, he says, "New York, which is so pretty during the holiday season," only to be replaced by "Laos" a few days later. In his mind, he travels the world back and forth at least twice before he is forced to make a decision immediately or lose his seat on the overbooked holiday-season flight. Last year, he eventually decided on Indonesia, while this year the debate is still going on. I am writing these words in early December, so there is still time for him to contemplate all 195 countries.

When he finally makes up his mind, after creating lists of pros and cons and reading multiple travel guides, something intriguing happens. Once the flight is booked and the tickets are issued, he becomes completely confident that he has made the best possible decision. This is days before he packs his suitcase and arrives at his destination. Uncertainty melts away and utter conviction takes its place. Indonesia is a great choice, he asserts; not only is the weather warm but the culture is fascinating and the experience novel. New York, on the other hand, as lovely a city as it is, will be painfully cold in January.

I have to admit that when it comes to my winter holidays, there is no debate—I go back home. Luckily, the weather there is beautiful year-round, and spending a week or two near the Mediterranean Sea leaves all other competing alternatives behind. I am not, however, resistant to indecision. It took me years (yes, years) to decide which job to take after completing graduate school. I went back and forth between U.S. universities on the East Coast and those on the West Coast, changing my mind multiple times for "the very last time," only to end up on a different continent altogether, in a country known for its queen, flavorless food, and rainy weather. Of course, I now think this is the absolute best path I could have taken, and it would be extremely difficult for you to convince me otherwise.

Coffeemakers, M&M's, and Vacations

Our tendency to reevaluate our options once we formulate a decision is a powerful one. After making a difficult choice between two equally valued options, such as between two desirable job offers or vacation destinations, people subsequently value the selected alternative more strongly than they initially had, and the discarded one less so. This phenomenon was first discovered by the psychologist Jack Brehm in 1956.[1] Brehm had recently gotten married, and on the spur of the moment, he decided to put his wedding gifts to good use in a unique way. He wanted to examine how the mere act of choosing changes our preferences. His new kitchen appliances and other household items were soon to make a historic contribution to our understanding of the human mind.

To make the task engaging, he had to recruit participants who would be interested in the items he had to offer, such as toasters, transistor radios, and coffeemakers. Brehm decided to recruit 1950s housemakers for his study. One bright morning, he invited a group of housewives to his lab. He presented them with the wedding gifts he'd recently received, then asked them to indicate how much they would like to have each of the items. How happy would they be with a set of wineglasses? A radio? A handheld mixer? They were then given a choice between two items they had rated similarly and were told they could take one item home. Which would they choose? The wineglasses or the radio? The toaster or the coffeemaker?

These were difficult decisions, but ones that had to be made, and so the women complied. Zelda selected the wineglasses over the radio, and Beatrice chose the toaster over the coffeemaker. After they had picked the items they wanted, Brehm politely asked his participants to rate all the items once more. How happy would they be with a set of wineglasses? A radio? A handheld mixer? Did making a decision change their preference?

It did. Each of the women subsequently affirmed that the appliance she had selected was even better than she'd initially thought, and the rejected option was not that great after all. Beatrice, who had chosen the toaster over the coffeemaker, now believed the toaster was much superior to the coffeemaker, whereas before making the choice, she'd thought the two were equal. Zelda now thought the radio was not as great as she'd originally indicated.

When the study was over, the women gathered their things to go home, happily discussing the best place in the kitchen for the brand-new toaster and which wine would be poured in the new glasses. Alas, it was confession time for Brehm. He was unable to let them take any of his gifts home. His wife would never forgive him if he did, he said, and he was not quite ready for divorce. The news was not received well by the participants.

Brehm may have become unpopular with the local female population; however, from that point on he was well recognized by psychologists around the world. His experiment, known as the *free-choice paradigm,* was replicated hundreds of times, and the idea that our actions alter our preferences has since been supported by large sets of data. Some of the most intriguing work in the field was done on very special participants—hairy creatures with a fondness for bananas and nuts, also known as capuchin monkeys. Laurie Santos, a professor of comparative and evolutionary psychology, runs a monkey lab at Yale University. Her aim is to explore whether aspects of human behavior that many of us consider uniquely human are, in fact, exclusive to humans, or whether they are rooted instead in our hairy ancestors.

To examine if monkeys, just like humans, change their valuation of items after making choices between them, she introduced the primates to a "food market" where they could exchange different items. Observing this food market, the researchers quickly learned the value capuchins assigned to different treats. A single Cheerio, for example, was worth five Rice Krispies. A marshmallow covered with fruit roll-up was considered the ultimate gourmet treat by the primates, and it would be traded only if the experimenter offered the monkeys a whole bowl of Cheerios in exchange. For a capuchin, a chocolate-covered banana treat was equivalent to a three-star Michelin meal for a human. A peanut could be thought of as an Italian gelato on a summer day. A sunflower seed was valued by the hairy ones as, say, a slice of pizza is by *Homo sapiens.* Using this system, the Yale researchers quantified the monkeys' preferences.

Think about a pack of M&M's—those small chocolate treats that come in all colors of the rainbow, with the letter *M* imprinted on them. Many of us have specific preferences when it comes to M&M's. Some like the brown ones better than the yellow ones, or the red ones better than the green. Some have no particular preference. If you do not care whether your M&M comes in blue, yellow, or green, you have something in common with capuchin monkeys. Using the food-market system, the researchers observed that capuchins valued the various colors of M&M's equally. The monkeys did not care if the chocolate was green, yellow, blue, or even purple—it was all the same to them. This was valuable information for the scientists. They would then examine whether choosing between two M&M's that were different in color changed the capuchins' preferences for the small round chocolate treats.

They made the monkeys choose between a yellow M&M and a blue one. Lo and behold, after making the agonizing choice (let's say a monkey finally settled on the yellow M&M), the monkey valued the rejected M&M (the blue one) less than it had before the decision was made.[2] Just like humans, the monkeys adjusted their preferences to align with their actions. Whereas before the choice, the monkey had been indifferent, now it seemed to desire the yellow M&M more than the blue one. It would trade the yellow M&M only if offered a larger amount of Cheerios than had previously sufficed. If monkeys show choice-induced alteration in preference, that indicates one of two things: Either monkeys are capable of complex rationalization or, more likely, reevaluation is mediated by relatively automatic low-level processes—processes that do not depend on highly evolved cognitive mechanisms. Liking a toaster more than a coffeemaker after we have chosen it—while having been indifferent about which was better prior to making a choice—is an inclination we inherited from our closest relatives.

Did the monkeys remember which color M&M they had chosen? Is that critical for preference change? If Tim suddenly suffered amnesia and forgot he had chosen Indonesia for his vacation destination, would he still value it more than all the options he rejected?

In 2001, a group of Harvard psychologists set out to examine whether amnesiacs show changes in preferences after making decisions, even though they cannot remember which option they have chosen.[3] The amnesiac patients had suffered hippocampal damage, which prevented them from being able to form new memories. As you probably remember, the hippocampus is a brain structure in the medial temporal lobe that is important for the formation and consolidation of memories that can be consciously retrieved. Patients with hippocampal damage can hold some information in their minds for a few minutes, but once they are distracted, that information is gone, never to return.

Rather than offering M&M's or household items, the Harvard psychologists presented the amnesiac patients with posters of abstract paintings. By now, you are familiar with the paradigm. Patients were asked to rank the paintings from the most preferred to the least preferred; then they were told to choose between two paintings they had ranked similarly. The patients made their choices, and then the researchers left the room. Thirty minutes later, they returned. At that point, the amnesiac patients did not recognize the researchers; they did not even recall participating in the study only half an hour earlier. Needless to say, they had no idea which poster they had chosen and which they had passed on. However, when given all the posters to rank in order again, the patients ranked the poster they had chosen higher than they initially had, without even knowing they had chosen it, and the rejected one lower. This means we do not need to consciously remember that we made a choice in order for that choice to change our preferences.

These demonstrations give us clues to the brain mechanisms involved in the phenomenon Brehm originally discovered. First, we do not need our hippocampi for our choices to change our preferences. Second, we know from the M&M's experiment with the Yale monkeys that the process relies on brain structures that are evolutionarily old. Where in the brain do these changes take place? Does the act of choosing actually modulate the neural representation of a stimulus's value? Do we really prefer the yellow M&M over the blue one after choosing it, or are we just fooling ourselves? Do we say we like the toaster better than the coffeemaker in order to seem consistent and feel good about our choices? Alternatively, does our emotional response to the toaster actually change after we pick it, maybe forever? My colleagues and I turned to the fMRI scanner for answers.

Heavily influenced by Tim's annual holiday decision-making anguish, I wanted to find out what happens in the brain when we contemplate possible vacation destinations, make a decision, and then reevaluate the options. My colleagues—the world-renowned neuroscientist Ray Dolan and the rising star Benadetto De Martino—and I adapted Brehm's classic free-choice paradigm to neuroimaging.[4] Our design was simple. We asked subjects to imagine going on vacation to eighty different destinations (such as Thailand, Greece, Florida, and Rome) and rate how happy they thought they would be if they vacationed in those places.

Imagine yourself in Paris; take your time. How happy would you be vacationing in Paris on a scale of 1 (not very happy) to 6 (extremely happy)? Next, imagine yourself in Brazil; really create a complete and detailed image. How happy would you be vacationing in Brazil? This is what our participants did for about forty-five minutes while their brains were being scanned.

You can guess what we did next. Yes, we presented the participants with two destinations they had rated the same and asked them to choose which one they would rather vacation in if they had to decide between them. Would you choose Paris over Brazil? Thailand over Greece? Finally, we asked them to imagine and rate all the destinations again. As expected, participants rated destinations they had selected higher after choosing them relative to before, and destinations they had discarded lower. The crucial question was, How were these changes reflected in the brain?

Our data revealed that the change was observed in the same part of the brain that responds to rewards such as food, love, or money—the caudate nucleus. The caudate nucleus, a cluster of nerve cells deep in the brain, is part of a larger structure, the striatum. The caudate has been shown to process rewards and signal the expectation of them.[5] If we believe we are about to be given a juicy steak or a hundred-dollar bill, or are about to engage in a sexual act, our caudate indicates this expectation. Think of the caudate as an announcer broadcasting to other parts of the brain information that has just been received from below: "Be ready for something good." After we receive our dinner or our monthly salary, the representations of these stimuli are quickly updated according to the value of the received reward. If we expect to get an allowance of $100 but end up getting $110 instead, the new, larger value of our allowance will be reflected in striatal activity. If the steak turns out to be a bit dry, the decreased value will be tracked, so that the next time we are about to eat steak, our expectations will not be as high as before.

The neuroimaging results showed that while participants were imagining future holidays, the activity in the caudate nucleus correlated with their expectations of how good they would feel if they vacationed in the various destinations. When they imagined going to a highly desirable destination, such as Greece or Thailand, the signal in the caudate was greater than when they imagined a less desirable destination, such as Sheffield or Ohio (no offense intended). After a decision was made, the caudate nucleus rapidly updated a signal that represented expected pleasure. If initially the caudate was announcing "thinking of something great" while Greece and Thailand were being imagined, after Greece had been chosen over Thailand, it was broadcasting "imagining something remarkable!" for Greece and merely "thinking of something good" for Thailand. This seems to suggest that our true hedonic response to a stimulus is changed by a simple commitment to it.

The Power of Agency

The interesting point is that in order for us to value something more after committing to it, we have to be the ones making the decision. If someone else makes the choice for us, the change in value is not observed. For example, remember the study involving the monkeys. If the researcher first presented the monkey with a yellow M&M and a blue one and then gave the monkey the yellow M&M, the monkey did not value the yellow one more than before or the blue one less. In other words, the monkeys did not reevaluate their options postchoice if they were not the ones making the decision. However, if the researchers tricked the monkeys into believing that they had made the choice when they actually had not, the monkeys showed the same reevaluation tendency.[6]

How did the clever Yale researchers trick the monkeys into believing they were making choices when actually they were not? They showed the monkeys a yellow M&M and a blue one. Then they pretended to put both in an opaque box and had the monkeys put their arms in the box and pick an M&M without looking. Unbeknownst to the monkeys, the researchers did not actually place both colored M&M's in the box. They put in only one type of M&M (let's say the blue one). So while the monkeys believed they were choosing between the yellow M&M's and the blue ones, they had no choice but to grab the blue ones. Nevertheless, once the monkeys took out the blue M&M's from the box, they preferred them over the yellow ones.

You might say, Yes, monkeys are, well, monkeys, but surely humans cannot be tricked into believing they prefer something that they think they chose when, in fact, they did not choose it? Or can they?

My student Cristina Velasquez and I decided to see if we could make intelligent students at University College London prefer one vacation destination over another simply by making them think they had picked it when, in fact, our computer program had randomly picked it for them.[7] First we had volunteers rate how happy they would be if they were to vacation in certain locations. Then we paired up equally rated vacation destinations and told volunteers they had to choose between them. However—and here comes the tricky bit—we told our participants that we would be presenting the two options subliminally. We said that the options would be presented very fast, only for a few milliseconds, and that they would be masked by a string of random symbols. So instead of seeing something like "Greece/Thailand" on the screen, the volunteers would see something like "*%%^/***&^" and would have to choose between the first option and the second. We told our volunteers that although the presentation of the options would be masked and presented too fast to process consciously, they should be able to pick up on the alternatives subconsciously, and thus make a decision based on preference.

This was a lie. We did not really present the options, not even for a few milliseconds; we presented only the random symbols. After a subject made his choice (let's say he picked the first option in response to "*%%^/***&^"), we revealed the true identities of the options: "Greece/Thailand." Now the volunteer believed he had chosen Greece over Thailand when, in fact, the choice had been random.

Nevertheless, after making these choices, the participants rated the "chosen" option higher than they had before the decision-making stage. Participants now indicated they believed they would be happier vacationing in Greece than in Thailand, whereas before the two options had been rated equally. Again, the change in ratings was observed only if the participants believed they were making the choices themselves. If we told them that the computer was making the choices for them, they did not reevaluate their options postchoice.

What these experiments tell us is that when you select something—even if it is a hypothetical choice, even if it is something you already have, even if you had not really picked it but just believe you did—you will value it more. From the retail industry to the workplace to our personal lives, the implications of this phenomenon are extraordinary. Imagine this scenario: A talented employee is approached by a competing company, which tries to recruit him away from his current position. He gets a very nice offer, which he might consider for a while, but ultimately decides to stay at his current workplace. My educated guess is that although the employee has decided to stay in the same position, with the same salary, the same benefits, and the same coworkers, and nothing has objectively changed, he will now value his job slightly higher than he did before, simply because he has chosen it again.

Take another example: My local coffee shop runs a special promotion. If you buy your coffee before 11:00 a.m., you get a piece of fruit or a croissant free of charge—your choice. Driven a bit by guilt, and by the fact that oddly enough I am not a big croissant fan (offer me a brownie and I believe my choice might be different), I go for an apple. I am quite happy with my free apple every morning. I choose a green Granny Smith. Would I be as happy if the store were offering only the croissant? Probably not. That's not so surprising, though, as I just told you I prefer apples. What if they were offering just apples—no choice granted? Would I be as happy with my free apple if I were simply given it? I suspect I would not be quite as satisfied if I were given an apple and had no choice. I am more content with my apple after choosing it over the croissant, the banana, and the orange. In both cases, I end up with the same apple, but the fact that I had debated the alternatives and selected the apple makes it slightly tastier.

The conclusion? If you would like to increase your employees' commitment to your company, your students' commitment to their studies, your clients' appreciation of the service you are providing, remind them every so often of their freedom of choice. Remind them of their decision to work at this company, to study at their selected college, and to use the provided services. An airline I often fly with does just that. At the end of every flight, the pilot says over the loudspeaker, "We know you have many options when making your travel plans and we thank you for choosing us. We hope to see you again on another flight." I am instantly convinced that since I chose this airline, it must be better than the rest and that I should probably make my bookings with the same airline again.

Has the institution of marriage remained so popular today because people value a partner more after officially committing to him or her? The same person, whom you have known and possibly lived with for many years, may seem a tiny bit more lovable after you officially choose to spend the rest of your life with him. I do not have data on this specific issue, but if you had people rate their partners before getting engaged and then again just after getting married, I predict ratings would go up. Of course, these ratings might eventually decline with time; otherwise, divorce rates would not be as high as they are. This decline could, however, occur because as years go by both partners in a marriage change. For example, a woman may sometimes feel the person beside her is no longer the one she chose years ago. She may also feel she is no longer the same person who made the original choice. As time passes, the sense of ownership of the decision may evaporate, and a need to make a new one may emerge.

Dr. H. Wallace Goddard, a relationship expert, seems to agree: "Strengthening and maintaining commitment in marriage involves a daily and continuing choice in which we choose the same marriage partner over and over again."[8] Dr. Goddard does not offer a reason why believing we have chosen our partner again every morning will make us happier in our relationship. You, however, can probably guess. Every day we make this decision, even if it is somewhat hypothetical, the value of our partner will slightly increase.

Why Decisions Alter Preference

Why is it that we value things more after selecting them? What actually triggers the tendency to reevaluate alternatives after making a decision? How is it that before buying a new car, or new shoes, we can literally spend hours in the store evaluating the options, going back and forth between the two we like best. However, once we leave the store with our new purchase, we feel genuinely content. We really and truly believe our new red pumps are much more suitable for our needs than the black ones, or that the Mini Cooper will make us much happier than the family car.

To explain this, the psychologist Leon Festinger introduced what was to become one of the most prominent theories in psychology—*cognitive dissonance theory.*[9] According to this theory, having to make a choice between two similarly desirable alternatives triggers psychological discomfort. This is because the decision conflicts with the desirable aspects of the rejected alternative, and with the undesirable aspects of the selected alternative. If you make up your mind to purchase a Mini Cooper over a family car, that decision clashes with the fact that the family car has room for your kids, while the Mini Cooper does not. According to cognitive dissonance theory, by reevaluating the options postchoice, in a way that is consistent with your decision, reduces psychological tension. So after sealing the deal on the Mini Cooper, you may say to yourself that the car will make you feel young, that it will be easy to park in the city, and that it is environmentally friendly compared with the large family car.

There are competing hypotheses to cognitive dissonance theory. The chief one is known as *self-perception theory.*[10] According to this theory, people infer their preferences by observing their choices. In other words, I can conclude that if I purchased the red pumps, that must mean I prefer them to the black pumps. I might not even remember why I settled on the flashy shoes. Nevertheless, the fact that I just spent a nice sum of money on them must mean I like them a lot. From the fact that the other pair is still on the shelf, I conclude that I did not want them as much. The result? My evaluation of the chosen red pumps is boosted and that of the rejected black pumps is lowered.

A key difference between self-perception theory and cognitive dissonance theory is that according to the latter, a feeling of negative arousal is key in driving preference changes. On the other hand, self-perception theory asserts that the negative feeling is not necessary. This crucial difference means that psychologists can directly test the two theories by manipulating the feeling of negative arousal. It turns out that when physiological arousal is not produced during decision making, preference changes are not observed. That's not all. When negative arousal is produced but people misattribute it to something else in their environment rather than to the decision-making process, reevaluation of the options does not occur. For example, when subjects were given a pill just before making a difficult decision and told that the pill might make them feel ill at ease (in fact, the pill was just vitamin C), they did not change their preferences postchoice. Although the participants felt a negative psychological arousal when making a tough decision, they mistakenly assumed that the arousal was due to the pill. Thus, they had no need to reduce the negative arousal by changing their preferences.[11]

There is another potential reason for the phenomenon. If I was offered a vacation in Brazil and one in Sicily, I would be extremely excited by either, rating each high on the desirability scale without dwelling on it for too long. However, if I had to choose between them, I would be forced to think about the alternatives in more detail. Before, a vacation in Brazil would simply make me think of relaxation and sunshine. After I was asked to decide between Brazil and Sicily, however, my thoughts regarding a vacation in Brazil would be "sunshine, but somewhat difficult to get to," while Sicily would be "sunshine and only a few hours' flight from London!" Making a difficult decision causes people to think a little more about the advantages and disadvantages of the alternatives. It highlights the unique aspects of the options (such as travel time in the case of Brazil and Sicily), which may not have been considered thoroughly before.

Predicting and Manipulating Choices

But are two options ever really equal to begin with? Or do small preexisting differences in preference ultimately tip a decision one way or the other? You may know what I mean if you have ever spent sleepless nights contemplating an important decision (moving to a new city, taking a new job, getting married, getting divorced), ultimately making the decision that you somehow knew all along you would eventually make. Many times we have an intuition about which path we will end up taking—a gut feeling, if you will. However, when the stakes are high, we often believe we should consider our options carefully before acting. By doing so, we try to increase our confidence in our choices. In some cases, the values of the options are so similar that these differences in preference are not even consciously accessible. Could brain activity tell them apart?

Although my colleagues and I did not set out to answer this question, that was exactly what we discovered when we looked at the brain-imaging data we acquired. We found that when our volunteers were imagining different vacations, their brain activity predicted which destination they would later choose. This was before they even knew they would be asked to make choices. For example, although Mary, one of our subjects, indicated that she felt Greece and Thailand were equally desirable, her caudate nucleus was telling a different story. Activity in her caudate was slightly higher when she imagined vacationing in Thailand than when she imagined going to Greece, suggesting her preference for Thailand over Greece. Indeed, when the time came to make a decision, Mary chose Thailand. After she made this decision, her slight preference for Thailand became greater.

What does this all mean? The findings suggest that when forced to choose between alternatives that we say we value equally, the decision is, in fact, not arbitrary at all. Our choices are determined by a difference in value that may not always be expressed in words but that can be detected by looking at our brain activity.

Before you get too excited and run to the nearest fMRI center to ask your caudate whether you should buy the blue scooter or the yellow one, take that job offer in Florida, or have meatballs or pasta for dinner (or both), I will have to curb your enthusiasm—fMRI technology is not able to read your mind.

What our findings show is that, on average, activity recorded by an fMRI scanner can indicate with a precision that is higher than chance which choices participants are likely to make. However, this is possible only when averaging over many decisions, and over activity acquired from a number of participants. The signal-to-noise ratio from the magnet does not permit accurate predictions on a trial-by-trial basis. For now, and probably for many years to come, you will need to dig deep within yourself for the answers.

Still, if we can predict *on average* the choices people will make by recording activity in the caudate nucleus while they contemplate the options, could we (on average) change their decisions by altering this signal? Could we modify their expected pleasure from the alternatives? Simple logic suggests we should be able to. My colleagues Tamara Shiner, Ray Dolan, and I decided to try to manipulate our volunteers' expectations by modulating brain activity while they contemplated their holiday plans.[12] There are a few different ways to alter neural processes. One could use deep brain stimulation. This involves implanting a "brain pacemaker," which sends electrical impulses to particular brain structures, changing their activity in a controlled manner. This method has been used to help patients with Parkinson's disease, as well as people with chronic pain, and has recently been shown to help in the treatment of depression (as described in chapter 6). Another method is transcranial magnetic stimulation (TMS), in which weak electric currents are induced in brain tissue by rapidly changing magnetic fields in a noninvasive manner. We, however, opted for an old-fashioned method. We would not implant a brain pacemaker or induce electric currents; we would use the more traditional pharmacological manipulation.

From our brain-imaging study, we knew that activity in the caudate nucleus during the process of imagining specific alternatives tracked expectations of pleasure and predicted the choices subjects would subsequently make. We believed that this activity was reflecting dopamine function, as dopaminergic inputs are dense in the caudate nucleus. Dopamine is a neurotransmitter that is required for learning about and processing different types of rewards, including food, sex, and money.

In order to manipulate activity in the caudate, we would alter dopamine function in our participants' brains while they thought about vacation alternatives. We did so by giving our participants L-dopa, a naturally occurring amino acid found in food that is converted into dopamine in the brain. It is often given to Parkinson's patients because they have low levels of dopamine as a result of their disease. When low doses are used (like the ones we administered), there are no significant side effects to taking L-dopa. In the United States, herbal supplements containing L-dopa are available without prescription.

When our volunteers arrived in the lab, the first thing we did was ask them to rate how happy they would be if they were to vacation in eighty different destinations. We then asked them to imagine vacationing at half of the destinations after the administration of L-dopa and half after the administration of a placebo (we used vitamin C as the placebo). The participants did not know which of the pills was which. We then sent them home and asked them to come back the next day. When they returned twenty-four hours later, we asked them to choose between pairs of destinations they had originally rated the same, then requested they rate all destinations again.

Would L-dopa change the volunteers' expected pleasure from the imagined vacations? It did. Participants rated destinations they had imagined under L-dopa higher after the manipulation relative to before. For example, if they rated Rome as a 5 ("will make me very happy") when they first came into the lab, then imagined going to Rome under the influence of L-dopa, they were more likely to rate Rome as a 6 ("will make me extremely happy") the next day. Not surprisingly, ratings of vacation destinations that they had imagined under vitamin C did not change from the first day to the second.

Would subjects also be more likely to choose the destinations they imagined under the influence of L-dopa over those they imagined under the influence of vitamin C? Although this effect was not very large, the answer was yes. The majority of the participants (67 percent) selected more destinations that they imagined under the influence of L-dopa than ones they imagined under the influence of a placebo. As L-dopa had enhanced the expected pleasure of the vacations, participants were more inclined to pick them over the others.

Modern society presents us with more choices than ever before. Unlike our ancestors, many of us can select from a near-infinite number of possibilities of where to live, whom to marry, which profession to embark upon, what to eat, and how to spend our leisure time. Neurons in the caudate that are sensitive to dopamine signal the predicted value of different options. By tapping into these signals, we can learn about the choices that people are likely to make at a later time.

After making a choice, the decision ultimately changes our estimated pleasure, enhancing the expected pleasure from the selected option and decreasing the expected pleasure from the rejected option. If we were not inclined to update the value of our options rapidly so that they concur with our choices, we would likely second-guess ourselves to the point of insanity. We would ask ourselves again and again whether we should have chosen Greece over Thailand, the toaster over the coffeemaker, and Jenny over Michele. Consistently second-guessing ourselves would interfere with our daily functioning and promote a negative effect. We would feel anxious and confused, regretful and sad. Have we done the right thing? Should we change our mind? These thoughts would result in a permanent halt. We would find ourselves—literally—stuck, overcome by indecision and unable to move forward. On the other hand, reevaluating our alternatives after making a decision increases our commitment to the action taken and keeps us moving forward.

After a long and difficult debate, Tim finally decided to spend his Christmas holiday in Costa Rica. He can already imagine himself relaxing on the magnificent sandy beaches, encountering M&M-loving monkeys in the jungle, and catching waves in the Pacific Ocean. These images trigger a burst of firing from dopaminergic neurons in his caudate. Yes, he expects to have a great time.

CHAPTER 9

Are Memories of 9/11 as Accurate as They Seem?

How Emotion Changes Our Past

On Friday, April 14, 1865, *Our American Cousin,* a comedy in three acts by Tom Taylor, was playing at Ford's Theatre in Washington, D.C. The play tells the story of a young American who crosses the ocean to claim his heritage from rich British relatives. In attendance that evening were Maj. Henry Rathbone, an officer and diplomat, and his fiancée, Clara Harris. Sitting beside them were Mary Todd Lincoln and her husband, President Abraham Lincoln.

All were well dressed for the occasion—the ladies in elegant frocks, the gentlemen in distinguished suits. None could have predicted the tragedies that were to come. For the major, insanity was in store; for Clara, a violent death brought about by her future husband. While Clara and Henry had a couple of decades before madness and murder would unfold, for Mary and Abraham this night would be their last together. At 10:15 p.m., John Wilkes Booth stepped into the presidential box at Ford's Theatre and shot Mary's husband. The laughter of the theatergoers, who were clearly enjoying the onstage comedy, was interrupted by her screams. President Abraham Lincoln was severely wounded. He died the next morning.[1]

Had the assassination taken place during the twenty-first century, cell phones would have been pulled out immediately to document the incident. Within the hour, images of the chaos inside the theater would have aired on Fox News and CNN. The world would have had a clear image of the event that very day. Back then, however, without the Internet, TV, texting, faxes, or even radio broadcasts, news traveled slowly.

> My father and I were on the road to Augusta in the State of Maine to purchase the "fixings" needed for my graduation. When we were driving down a steep hill into the city we felt that something was wrong. Everybody looked so sad, and there was such terrible excitement that my father stopped his horse, and leaning from the carriage called: "What is it, my friends? What has happened?" "Haven't you heard?" was their reply—"Lincoln has been assassinated." The lines fell from my father's limp hands, and with tears streaming from his eyes he sat as one bereft of motion. We were far from home, and much must be done, so he rallied after a time, and we finished our work as well as our heavy hearts would allow.

This quote is taken from the very first investigation of what we now term *flashbulb memories*. It was published in 1899 in the *American Psychologist* by a scientist named F. W. Colgrove.[2] In his paper, entitled "Individual Memories," Colgrove described people's memories of learning about President Lincoln's assassination. Colgrove found that most people could recall in amazing detail what they were doing and where they were when they heard President Lincoln was assassinated, even years after the event took place. The examples he provided tell a story of a different time and place, but the human experience is extremely familiar.

Fast-forward to Friday, November 22, 1963, Dallas, Texas. In the presidential limousine heading toward Dealey Plaza sat John Bowden Connally, Jr., the governor of Texas, and his wife, Nellie. Behind them were Jacqueline Kennedy and her husband,

President John F. Kennedy. It would be their last ride together. At 12:30 p.m., Lee Harvey Oswald shot Jacqueline's husband from a nearby building. The cheers of those in the crowd, excited at the sight of their president and First Lady, were interrupted by her screams. Thirty minutes later, President John F. Kennedy was pronounced dead.

Images of Kennedy's assassination were recorded and documented. Photographers and cameramen who were following the president on his Texas tour captured Kennedy's last moments. As a result, the investigation of the incident did not rely solely on the testimony of eyewitnesses. The fatal car ride, however, was not broadcast live. It took a few days before footage of the incident was first aired, and even then the broadcast was local. The most complete visual recording of the assassination, taken by Abraham Zapruder, was shown on television only years later.

In what became a seminal paper in the field, two Harvard psychologists, Roger Brown and James Kulik, examined people's memories of learning of the assassination of John F. Kennedy.[3] Like Colgrove, they found people's memories of learning about the incident to be extremely detailed and vivid. They noticed that people's recollections of these shocking events usually included answers to the following questions (see if you can easily answer these same questions regarding your experience on September 11, 2001): "Where were you?" "What were you doing?" "Who told you; how did you find out?" "What were the feelings of others around you?" "What were your own emotions?" "What was the aftermath?"

On the basis of their investigation, they suggested that the surprising and consequential nature of these public events triggered a unique mechanism that preserved what had occurred at that instant, producing a picturelike representation that they termed a *flashbulb memory.* Brown and Kulik assumed that these vivid and detailed recollections were unusually accurate. However, they had no way of assessing the actual validity of these memories, and it later turned out they were completely wrong.

Their study rested on an analysis of memories reported several years after JFK's death. They did not have the data needed to validate these memories. It wasn't until Ulric Neisser, a Cornell professor and a member of the National Academy of Sciences, compared flashbulb memories with self-reports collected shortly after the shocking events that the truth surfaced.

On January 28, 1986, the space shuttle *Challenger* exploded in midair only seventy-three seconds after liftoff. The launch was aired live on CNN, as the crew included the New Hampshire schoolteacher Christa McAuliffe. McAuliffe was the first teacher to be sent into space as part of the Teachers in Space project. NASA had arranged for children in public schools to watch the launch live on television, resulting in the live viewing of the explosion by thousands of schoolkids. The disturbing pictures of the disaster were replayed throughout the day on network TV. Because there was widespread media coverage of the incident, 85 percent of Americans learned of the explosion within an hour of the accident.

Less than twenty-four hours later, Neisser began his investigation of flashbulb memories of the explosion. He interviewed college students and asked them where they were and what they were doing when they learned about the incident. He then did the same thirty months later. What he now held was something Brown and Kulik never possessed. He had the data needed to examine the accuracy and consistency of flashbulb memories. By comparing people's initial accounts of how they learned about the *Challenger* explosion with their memories two and a half years later, he could empirically test whether these memories were exceptionally resistant to being forgotten, or whether they just appeared to be.

His findings were astonishing. Twenty-five percent of the

respondents were mistaken about every single detail of how they had learned of the disaster. There was absolutely no match between their later memories of how they had learned about the explosion and how they actually had. Take this participant's initial account of hearing about the *Challenger* explosion:

> I was in religion class and some people walked in and started talking about [it]. I didn't know any details except that it had exploded and the schoolteacher's students had all been watching which I thought was so sad. Then after class I went to my room and watched the TV program talking about it and I got all the details from that.

Thirty months later, this is how the same participant recalled hearing about the explosion:

> When I first heard about the explosion I was sitting in my freshman dorm room with my roommate and we were watching TV. It came on a news flash and we were both totally shocked. I was really upset and I went upstairs to talk to a friend of mine and then I called my parents.[4]

Not everyone did this badly. Half of the students were incorrect in about two-thirds of what they recalled. A mere 7 percent of respondents received perfect scores—their memories of the *Challenger* explosion thirty months later and their initial accounts were identical. What was even more surprising was that almost all of the participants were certain that they remembered the events exactly as they had unfolded. On a scale from 1 (not being confident at all in the accuracy of the memory) to 5 (being 100 percent confident that the memory was an exact representation of what had occurred), the mean confidence rating of the participants was a whopping 4.17. In other words, the students were pretty sure they were providing precise recollections. Furthermore, there was absolutely no correlation between the accuracy of the memory and the confidence in which it was held. This means that, in many cases, people were certain about their recollections when, in fact, these memories were entirely false.

Neisser's groundbreaking study showed that flashbulb memories are not so much Polaroid photos as snapshots brushed up in Photoshop again and again. The retouched photo might resemble the original image, but it is no longer an exact representation of what was initially captured. While Neisser's results clearly show that flashbulb memories should not be taken as exact replicas of the event they portray, the question still remains: Are they better at representing the original occurrences than memories of mundane, everyday events? Although not completely accurate, do we remember the events of 9/11 better than dinner last night?

Being There Matters

On that Tuesday morning early in September, I got up lazily and made some coffee. I had about an hour before my first class. From the only window in my tiny loft on Sixteenth Street, I could see it was a lovely day. A couple of minutes later, a friend of mine called. He was already at his office in midtown. Apparently, a plane had crashed into the World Trade Center. I turned on the TV to see what was happening. On the *Today* show images of a smoky tower were airing. It was unclear what had just happened. They were speculating that a small plane had accidentally crashed into the North Tower. On the other end of the line, my friend, who was a licensed pilot, claimed this could not be so.

"It is impossible to crash into a two-hundred-square-foot building that is one thousand three hundred and sixty-two feet high by mistake," he said, "and certainly not on a day with clear visibility like today."

What transpired in the next hour or so is somewhat blurry. I can just assume I continued to follow the events on TV while the second plane flew straight into the South Tower. The next thing I remember is watching, horrified, on live television as the South Tower collapsed. After about twenty minutes, not knowing what to do, I ventured out into the street.

Obviously, I was not at all ready for what awaited me downstairs. Masses of people were walking north along Sixth Avenue, away from the towers. Many of them were covered in dust, still wearing business suits, which just a couple of hours ago had been clean—men carrying briefcases, women in fancy high-heeled shoes. Long queues were forming near public phone booths, as mobile phones were no longer working (landlines would soon stop functioning, as well). And then we all watched as the North Tower went down.

I was utterly surprised by the collapse of the second tower. I had watched the first tower disappear about thirty minutes earlier on TV; I also knew that both towers had been hit by commercial planes in a similar fashion. You do not need to be a trained scientist to put two and two together; the second tower would most probably collapse, as well. However, I was unable to forecast this predictable turn of events. (Was this just another example of the human tendency not to believe in the worst-case scenario? Or maybe confusion took over.) In fact, the cloud of debris spread so far that I wasn't even sure what it was I was witnessing. Did the second tower collapse? Or was it a nearby apartment building? It seemed like the structure falling was only a few feet away, though I was more than two miles away.

Many salient events had taken place throughout my years in New York City, but that brief moment at 10:28 a.m on September 11, 2001, stands out in particular in my mind: the tower collapsing, the people around me crying out in shock, the man on my right-hand side, the lady across the street in a purple dress, the clouds of dust, and the warm sun. In the words of the "godfather" of experimental psychology, William James, "An impression may be so exciting emotionally almost to leave a *scar* upon the cerebral tissues."[5] This seemed to be the case. Or was it?

Although you could easily convince me that my memories of last Wednesday are filled with faulty details, you will have a very hard time persuading me that my recollections of 9/11, now years old, are inaccurate. Nevertheless, I am about to suggest that they may be.

On September 12, 2001, the psychologists Jennifer Talarico and David Rubin recruited fifty-four students from Duke University and asked them to write down their experiences of hearing of the terrorist attacks. That was not all; they also asked them to describe everything they had done the day before the attacks, Sepember 10, 2001. This provided them with what psychologists call a "control condition"—a baseline against which they could compare the rate at which flashbulb memories are forgotten. On September 10, 2001, the day before the attacks, most students experienced a normal, uneventful day. They did what students normally do on a Monday—they went to class, studied in the library, did their laundry, and had drinks with friends.

Some participants were invited back to the lab a week after giving the initial accounts so that their memories could be tested. Others were asked to return either forty-two days later or seven and a half months later. They were asked to write down everything they remembered from September 11, 2001, and everything from the day before, September 10. Would their memories of everyday events be different from their memories of 9/11? Would they forget them faster? The answer was yes . . . and no.

Talarico and Rubin found that memories of learning of the terrorist attacks on September 11 were forgotten at the exact same rate as memories of normal, everyday events.[6] While some details were remembered accurately months after the events took place, some were forgotten and others were remembered inaccurately. Overall, students were no better at recalling September 11, 2001, than they were at recalling September 10, 2001. However, there was an important difference between memories of the terrorist attacks and those of doing the laundry or going to class. It was not the objective accuracy of the memory; it was the *subjective* qualities of the memory.

Students were much more likely to believe that the events of 9/11 occurred exactly the way they remembered them, and were unlikely to be persuaded otherwise. Not only was their confidence in their memories of 9/11 higher than it was of their memories of 9/10; they also had more vivid memories of the terrorist attacks than of other events. They said they were more likely to feel as though they were reliving the experience all over again, and felt as if they were traveling back in time to September 11, 2001. They did not have the same experience when recollecting going to class or to the gym the day before.

Talarico and Rubin reached very similar conclusions to those arrived at by Neisser more than a decade earlier: Flashbulb memories are not more accurate than "regular" memories, but they certainly seem so. How could this be? For recollections of neutral events, the accuracy of a memory and our confidence in it usually go hand in hand. Why is it that when it comes to memories of highly emotional events, such as 9/11, the explosion of a space shuttle, or the assassination of a president, our confidence in our memories are no longer a suitable indication of their validity? To answer this question, the functioning human brain needed to be examined.

As chance would have it, my colleagues and I were in an ideal position to do so. In 2001, I was conducting research at New York University, specializing in the effects of emotion on memory. NYU is situated near Washington Square Park, in the center of Greenwich Village. Located in the heart of one of the most stimulating cities in the world, the Village has a neighborhood feel to it. It is also a mere two miles from Ground Zero.

On September 11, 2001, the department still did not have an fMRI scanner available for use. The scanner arrived about a year later, and approximately three years after that Tuesday morning in September 2001, we launched our investigation of the neural mechanisms mediating flashbulb memories. We wanted to find out if unique neural mechanisms were involved when people recalled personal events from 9/11, compared with their recollection of more mundane events. For our brain-imaging study, we recruited individuals who had been in Manhattan on the day of the attacks and asked them to recall their experience of 9/11 while we scanned their brains. Using brain imaging, we were able to see which parts of the brain were active when participants recalled those horrific events.

As had been the case with Talarico and Rubin, we, too, needed a baseline condition against which we could compare memories of 9/11. We chose to contrast recollections of 9/11 with recollections of the preceding summer. So in addition to recalling events of 9/11 while in the fMRI scanner, subjects were also asked to recall events from July and August 2001. Participants' memories of those months were often of events such as a summer internship, taking summer courses, or visiting a foreign country. These were unique and memorable events that our volunteers could retrieve three years after they had occurred, but they were not traumatic or highly surprising incidents such as 9/11.

Our participants were in the fMRI scanner for about an hour as they recalled all these events. Via a mirror, which was placed in the scanner, the participants could view a computer screen.

We presented word cues on the screen to help them come up with specific memories. In addition, either the word *September* or *summer* came up to indicate if they were to retrieve a memory from September 11 or one from the summer before. For example, if we presented the word *friend* together with the word *September,* the participant had to retrieve an autobiographical memory from 9/11 related to a friend. They retrieved about sixty memories. When they were finished, they came out of the scanner and sat in front of a computer. We asked them to recall all the memories again, but this time they were to type them out. We also asked them how vivid their memories were and how confident they were that they'd remembered the events exactly as they had happened. Did they feel as if they were reexperiencing the events while recalling them? How emotionally arousing were the memories?

We expected people's memories of 9/11 to be more vivid, more emotionally arousing, and held with greater confidence than memories of the summer before. However, our data told a different story.[7] Only half our participants indicated that their memories of 9/11 were more vivid, more emotionally arousing, and held with greater confidence than memories of the previous summer. For the other participants, memories of 9/11 were no different from memories of the summer before. What distinguished these two groups? Why did the 9/11 memories of half our participants possess the qualities of flashbulb memories, while those of the other half did not?

According to the Pew Research Center for the People & the Press, 51 percent of New Yorkers and 38 percent of Americans in general cited the terrorist attacks of 9/11 as the biggest event in their personal lives in 2001.[8] Obviously, you did not have to be in Manhattan, or even in the United States, on 9/11 to be able to recall that day. People around the globe have personal stories of their experiences of it—each of us has his or her own narrative of the day, and most of us have told it again and again. However, what we were about to discover was that when it came to the *subjective quality* of the memories, it did matter whether you were two miles from Ground Zero, as opposed to twenty thousand miles away, when American Airlines Flight 11 crashed into the North Tower of the World Trade Center. In fact, it even made a big difference if you were two miles as opposed to five miles away when it happened.

As part of the study, our subjects were asked to fill out a questionnaire regarding their personal experiences on 9/11. Among other questions, they had to indicate where exactly they had been on that day, if they had known anyone in the towers, and how the attacks had affected their personal lives. It turned out that what determined the subjective quality of memories of 9/11 was the exact distance people were from the World Trade Center at the time of the attacks.

For people who had been, on average, two miles from the World Trade Center (in downtown Manhattan), memories of their experiences on 9/11 were exceptionally vivid, and these individuals were very confident of their accuracy, much more so than of memories of the summer before. However, for those people who had been, on average, four and a half miles from the World Trade Center on 9/11 (around the Empire State Building), memories of that day did not feel much different from the control memories. Although all our volunteers had been in Manhattan on 9/11, the recollections of those who had been in downtown Manhattan, close to the World Trade Center, were qualitatively different from the recollections of those who had been farther away.

What differed in the experiences of these two groups that day that left "a scar upon the cerebral tissues" for one group but not the other? Only those who had been downtown could see the towers fall, hear the explosions, smell the smoke. "I saw with my own eyes: the towers burning in red flames, noises and cries of people," one volunteer said. Those who had been all the way down at Ground Zero had actually participated in the events. One of the most chilling accounts given in our study was by a man I will call Matt, who was working on Wall Street on 9/11.

> I remember getting out of the Wall Street subway station and pieces of paper falling from the sky all around me. I looked up and saw smoke rising above my building. I went up to my office when a coworker was talking about how he just saw a plane hit the WTC. We decided to go over and see what was happening and stood on the corner of Broadway and Liberty Street, in front of Liberty Park, looking at the enormous hole and tremendous flames engulfing the top of the building. While we were staring at that, the second plane flew into the South Tower. The explosion caused everyone in the area to automatically duck for cover, then turn and run. As the mass was running away from the burning towers and falling debris I remember an old lady getting knocked to the ground next to me and people trampling her. As I robotically ran across Broadway, a screeching car broke my spell of wild flight as it [braked] to avoid hitting a guy running across the street in front of me. That awoke me to the greater world outside of fleeing blindly from the explosion, and I saw some scaffolding that I could go under to avoid the falling debris and brace myself against the stampede of fleeing people. I waited there, watching the towers, until the scene calmed down a little, and then I made my way back to my office to tell anyone who might be there to get out. I also remember walking up Broadway through TriBeCa with my coworker toward my apartment to get away from the towers. I remember the giant hole in the side of the tower and the flames against the clear blue sky. I remember people in the streets crying and screaming and listening to the radio from cars in the street. I recall listening to the report of planes hitting the Pentagon from one of those cars while staring at the towers, when we began to see people jumping from the top

> floors. I watched as five or six people jumped, when I started to imagine what it must be like up there for them to want to jump off the 110th floor. I had to look away when I heard someone scream, and when I turned around, the building was falling to the ground in a plume of smoke and debris. I will never forget this.

One can only imagine the emotional impact of personally experiencing such events. It was quite clear that having been far away from the heart of it all produced memories that were very different: less emotional and less salient. People in midtown Manhattan had been too far away to see the plane crash or the tower fall. They learned of the events from friends or the media. "I was in the office and heard about the attack. I looked on the Internet," one subject said. Another recalled, "I remember watching TV news coverage at Caffé Taci [this was way uptown, near Columbia University] and presumably hearing sounds of explosions on TV."

So while participants who had been downtown on 9/11 had had direct personal experience of the terrorist attacks and reported feeling threatened, others were experiencing the events secondhand. Not only did people who had been closer to the towers feel that their memories were more vivid; they also used more words when describing them and they conveyed more details. One of the participants addressed this divergence in experience and its impact on her personal life: "It was frustrating trying to talk about it with my boyfriend back in California; there was just no way he could comprehend our different experiences of the event and therefore different outlooks. . . . We broke up shortly after."

I do not know if any of these memories is accurate. I do not have personal accounts taken on 9/11 against which to compare these recollections. Unlike Talarico and Rubin, I am unable to tell you whether and how these memories are different from memories of everyday events such as doing the laundry. I can, however, report that for people who were there—staring at the towers falling and at the victims jumping to their deaths—memories of those experiences were qualitatively different from memories of other past memorable events. In contrast, for individuals who learned of the towers falling via the Internet or TV, their memories, although vivid, were not very different from memories of a summer internship or a move to a new city.

Changes Observed in the Brain

We turned to the fMRI data we had collected to see how these differences were conveyed in the brain. Was a participant's distance from the falling towers on 9/11 expressed in a difference in brain activation three years later, when those events were recalled?

We identified two specific patterns of brain activation during recollection that could give us a clue as to whether someone had been at arm's length from the burning towers or more than a few miles away. First we observed clear changes in the activity of the amygdala. I have referred to this structure a few times in previous chapters. When the neural systems involved in mediating aspects of emotion were first considered seriously back in 1927, the amygdala was not recognized as having a key function. Its relevance to fear and anxiety was only suggested in the late 1930s when two researchers, Heinrich Klüver and Paul Bucy, reported that monkeys with lesions in the medial temporal lobe (where the amygdala sits) no longer seemed to be afraid of anything much.[9] But it wasn't until 1956 that the amygdala was identified as the site of the specific lesion that resulted in this emotional deficit.[10] Since then, the amygdala's role in processing emotion, and mediating the influence of emotion on memory, has been keenly studied and documented.[11]

Studies in animals show that the amygdala is especially important for expressing fear, as well as for learning about dangerous stimuli. For example, when rats face adverse situations, such as being shocked, they freeze. They are rapidly able to learn where they are likely to encounter the shocks, and if you give them the opportunity, they will avoid a chamber where they were previously shocked. However, if you cause a lesion in the amygdala, they no longer learn to avoid these dangerous places. They also do not express fear (i.e., they do not freeze) if they are put in a chamber in which they were previously shocked.[12] It seems the poor creatures just can't remember traumatic events without an intact amygdala, and thus fail to avoid danger.

When it comes to the neural circuits of emotional memories, we are a bit like rats. When faced with an emotionally arousing situation, such as a car accident or a physical attack, the human amygdala responds fiercely. Not only does it influence our immediate emotional reaction to the situation, but it is critical in affecting how the memories of these arousing events are stored for the long term. The amygdala modifies the storage of memories both directly, by projecting to other brain structures involved in the consolidation of memories, such as the nearby hippocampus, and indirectly, via stress hormones that enhance memory consolidation.

We can assume that the amygdala of people who were in close proximity to the World Trade Center when it collapsed responded more strongly than those of people who were sitting in their living rooms, watching the events on TV. Although simply hearing about shocking public events may result in arousal, the strength of this response likely varies depending on the individual's personal experience of the events.

New Yorkers who found themselves downtown on that day

were in what we call a "flight or fight" situation. When you encounter danger, such as an intruder at night or a bear in the woods, your body gets ready for action—your heartbeat is elevated; your breathing becomes quicker—and you either choose to flee the source of danger or fight it. The closer people were to the towers on 9/11, the greater the immediate danger to their own lives, and thus the greater their need to respond quickly. For people who were extremely close to the towers, like Matt, who participated in our fMRI study, the only possible response was to escape. We are all familiar with those photos of masses of people running away from the falling towers, trying to flee the huge cloud of debris. I suspect that the stress-hormone levels of these individuals were extremely high, maybe higher than they had ever been, and those elevated levels were probably maintained for quite some time.

Because the towers were especially tall and the cloud of debris vast, people who were a couple of miles from Ground Zero felt closer to the towers than they actually were. This was my own experience while watching the North Tower fall from about two miles away. As I noted earlier, the visibility was good that day, and the cloud of debris spread far; therefore, my perception was of a nearby building collapsing. I suspect my amygdala was signaling danger quite intensely, but almost certainly not as powerfully as Matt's.

Farther north, my friend, who was in his office in midtown, could see the smoke in the distance and hear the ambulances and fire trucks rushing downtown. He might have felt somewhat in jeopardy himself, but his brain was not indicating an urgent call to escape, or a need for immediate action of any kind. Although his stress-hormone levels may have been higher than usual, they were most likely nowhere close to those of people on Wall Street, or even to those of the lady standing across from me on Fourteenth Street.

All of this, however, is an educated guess on my part. I did not take blood samples of people across town on 9/11; neither did I record their amygdala activity that day. I did, however, record the activity of Matt's amygdala and that of twenty-two other New Yorkers three years later. Sure enough, when asked to recall their own experiences of the terrorist attacks, people who had been in downtown Manhattan on that day, such as Matt, showed greater activity in their amygdala than people who had been in midtown. The closer my participants had been to the WTC, the stronger their amygdala reacted when thinking about that day. The signal in the amygdala was directly related to how strong and vivid participants felt their memories of 9/11 were; the closer they had been to Ground Zero, the more emotional and vivid their memories were, and the greater the amygdala response during recollection.

Our brain-imaging data revealed another important clue to how people's distance from the WTC on 9/11 affected their recollections. When the downtown group thought about 9/11, they showed lower activity than normal in their parahippocampal cortex. This part of the brain is thought to be involved in processing and recognizing details of a visual scene. Psychologists have previously discovered that when we view an emotional event, our attention is focused on the central arousing aspects of the event (such as the towers collapsing) at the expense of peripheral details (such as the people standing next to us). The outcome is poor encoding of peripheral details, which results in less involvement of the posterior parahippocampal cortex during encoding and retrieval of memories. If neurons of the parahippocampal cortex are less active when arousing incidents are recollected and neurons of the amygdala are more active, this may explain why when we recall shocking events, we remember the central emotional details and our feelings at the time but cannot always provide accurate details about our surroundings.

When I recall standing on Sixth Avenue and watching the big cloud of dust approaching me rapidly, those feelings of confusion I experienced at the time are easily triggered, and I am quickly taken back in time. My emotional reaction during recollection creates a sense of a clear and vivid memory. The feeling that my memory is genuine may be true in part. I might indeed remember the tower falling and my emotional reaction to it accurately, details for which recollection may be mediated by the amygdala, but some other details, which depend more on the function of the parahippocampal cortex, such as the purple dress of the woman across the street, may be less reliable.

It is critical to understand precisely which details of emotional events are remembered better than those of mundane events, and which are remembered less well. Scientists are hard at work trying to solve this problem. While we still do not have clear answers, we do know that when it comes to the most arousing events of our lives, our confidence in our memories is not a reliable indication of how accurate they are. This has important implications for the legal system, especially regarding the validity of eyewitness testimony, which can often be inaccurate without any bad intention on the part of the witnesses.

Consider, for example, the death of Jean Charles de Menezes.[13] On July 22, 2005, de Menezes was shot dead by Metropolitan Police officers at the Stockwell Tube station in London. At first, witnesses said he had leaped over the turnstile, escaping from police. Soon after, it became clear that no such thing had occurred. De Menezes did not run from the police or jump over the barrier. The testimonies of eyewitnesses turned out to be inaccurate on many counts. Recollections of what de Menezes was wearing, exactly how the officers responded, and the number of shots fired at him were inconsistent. It later turned out that de Menezes had been mistakenly identified by the police as a suspect in the previous day's failed bombings, while in fact he was innocent. The story is a complicated one, as police failed to set the record straight and leaned on the wildly inaccurate reports of the witnesses. Eventually, the truth was leaked to the press, and the police were accused of falsifying information to aid their case.

The function of memory is to be able to use past experiences to guide future thoughts and actions. If an event is salient in our minds and we believe it to be true, we will act upon it regardless of its absolute validity. For example, if one night you are violently attacked while walking alone in the park, you are unlikely to enter a park unaccompanied when it is dark ever again. It does not matter if you accurately remember which part of the park it was, what the attacker looked like, or the exact time of the attack. The brain does not have the capacity to hold on to *all* pieces of information. However, it is crucial that you have a confident memory of the episode so that it serves as a constant reminder not to walk alone in secluded areas at night.

When it comes to our recollections, it is critical that we have vivid memories of the good, the bad, and the ugly—even if these do not provide perfect replicas of the events being recalled. A child needs to remember the terrible burning sensation felt when touching a hot oven so that he does not attempt to grab a muffin (or was it a bun?) from inside an oven again. A vivid memory of failing an exam drives us to study harder to ace the next one, and easily recalling past heartaches can help guide us in our next attempt at romance. Believing that we can use a negative past experience to learn and do better in the future may, in fact, fuel optimism. Optimistic people are not necessarily those with a positively biased view of the *past;* neither are they the ones holding a positively biased view of the *present.* They are the ones who see the *future* through rose-tinted glasses *despite* all the disappointing experiences they have had.

CHAPTER 10

Why Is Being a Cancer Survivor Better Than Winning the Tour de France?

How the Brain Turns Lead into Gold

Which would you rather be—the winner of the Tour de France or a cancer survivor? You probably don't need long to consider the options. Most likely, you are convinced I have lost my mind for even asking such a ridiculous question. Of course, as physically challenging as it may be, you would much rather be the renowned champion of the famous cycling race held in France every year, in which cyclists cover 2,175 miles in about twenty-three days. There is no doubt that none of us would choose to go through dreadful chemotherapy treatments—the reality for most cancer patients. The truth is, however, none of us is fully qualified to answer this question. This is because none of us has experienced both of these options. The best we can do is try to imagine what it would be like to be a cancer survivor and what it would be like to be the winner of the Tour de France. The former elicits images of hospital rooms, doctors, loss of hair, loss of weight, feelings of fatigue, nausea, pain, fear, and sadness. The latter elicits feelings of joy, excitement, achievement, fame, and happiness. How accurate are these predictions? There is only one man on this planet who can answer this question. He is Lance Armstrong, the seven-time winner of the Tour de France; he is also a cancer survivor. This is what he has to say:

> The truth is, if you asked me to choose between winning the Tour de France and cancer, I would choose cancer. Odd as it sounds, I would rather have the title of cancer survivor than winner of the Tour, because of what it has done for me as a human being, a man, a husband, a son, and a father.[1]

Lance Armstrong was born in Texas in 1971. At the age of twelve, he began participating in triathlons. It quickly became clear that his strongest event was cycling. Although he competed successfully as a rider in the early 1990s, he was considered unremarkable in comparison to the top cyclists of his time. Then, during the 1996 Tour de France, Armstrong suddenly became ill and dropped out. A few months later, he was diagnosed with testicular cancer. The cancer had already spread to his brain and lungs. He underwent surgery to remove the brain tumors and his diseased testicle. Less than two years after his diagnosis, Armstrong returned to professional cycling, stronger than ever. In 1999, he won his first Tour de France, and he continued to do so for six consecutive years.

Would Armstrong have achieved all this if his life had been a smooth ride, devoid of the challenges posed by cancer? Maybe. Or he might have remained a talented but unexceptional athlete. We will never know, and it does not really matter. What does matter is what Armstrong *believes* would have been. If you have read his autobiography, *It's Not About the Bike: My Journey Back to Life,* I suspect you would agree that Armstrong believes his fight with cancer gave him unexpected strengths and possibly a new outlook on life, one that enabled him to pursue and attain his personal and professional goals. So while people who have never had to endure cancer perceive the prospect solely in a negative light, Armstrong, along with other cancer survivors, sees the gain in what others view as misfortune.

The "Philosopher's Stone"

Our minds seem to possess the "philosopher's stone," which enables us to turn adversity into opportunity. In the ancient practice of alchemy, the philosopher's stone was believed to be the key element with which one could turn common metals into gold and silver and create a "panacea," a remedy that would cure all disease. For about 2,500 years, up until the twentieth century, philosophers and scientists from ancient Egypt to Rome to China devoted their lives to the search for the philosopher's stone. Despite an admirable recent attempt by Harry Potter and company, the stone that gives its owner eternal life was never found. As hard as they tried, the alchemists could never turn metal into much else.

The human brain, however, is extremely efficient in turning lead into gold. It does so rapidly, with what appears to be minimum effort. Our minds seek and adopt the most rewarding view of whatever situation befalls us. Although we dread hardships, such as divorce, unemployment, or sickness, believing that we will never get over them, we are usually wrong. People tend to bounce back to normal levels of well-being surprisingly fast following almost any misfortune. Merely a year after becoming paraplegic, accident victims report levels of enjoyment from everyday events similar to those of healthy individuals.[2] They also do not differ in the degree of future happiness they predict for themselves. Within a couple of years of getting divorced, people report the same level of satisfaction with life as they did a year prior to the divorce. Those who have been widowed take a bit longer to bounce back to normal levels of well-being, but they, too, climb back to baseline levels within a few years of their spouse's death.[3]

The irony, however, is that people are extremely bad at predicting how they would feel if they had to face such misfortunes. If you ask people to estimate how they would cope following the death of a loved one, or after becoming paraplegic, they tend to overestimate the length and intensity of their emotional reaction. The usual response is "My life would be over; I could not go on." You never hear someone say, "Well, if my husband divorces me, I will be back to my old self, feeling as happy as ever, in no time," or "If I lose the ability to use my legs, I will probably be as optimistic about the future as the next guy." In most cases, however, this becomes reality. With regard to a large range of medical conditions, patients report a significantly higher quality of life and pleasure than healthy individuals predict they would have if they suffered these conditions.[4]

Take Matt Hampson, for example. Matt is twenty-three years old. One day, during what seemed like just another rugby training session, Matt's life changed forever. In an unfortunate turn of events, he dislocated his spine and was paralyzed from the neck down, probably for the rest of his life. In a matter of seconds, Matt was transformed from a strong, independent young man to one in need of care around the clock. He now sits in a wheelchair, which he steers by using his chin, and he breathes through a ventilator. Most of us automatically feel pity for Matt. We dread finding ourselves in his shoes. Matt, on the other hand, says, "Life is different now. It's not over, it's different. And it's not any worse. Some ways it's better."[5] In some ways it's better because once Matt lost certain abilities, such as the ability to play rugby, he compensated by gaining new skills and exploring different capabilities. In his new life, Matt is writing a rugby commentary and an autobiography. He runs a rugby website as well as a charity for children with similar injuries. He is building a new house and coaches the local rugby team. I daresay that most of us pale in comparison.

The trick the brain plays once it encounters the unbearable is to quickly find the silver lining. Before we become severely ill, we view sickness and disability as something to be avoided at all costs. This is an adaptive way of viewing adversities, as it drives us to shun hardships, to keep away from danger, and to take care of ourselves. However, once these adversities become our reality, viewing them as such is no longer helpful. In order to continue functioning, we quickly need to reevaluate our circumstances and reverse our evaluation of the situation that has befallen us so that we can carry on with our lives.

The difference between believing life would not be worth living if we were in a wheelchair and the actual experience of people with disabilities living a full and satisfactory life is an example of a persistent error known as the *impact bias*. The impact bias is our tendency to overrate the effect of an adverse outcome on our well-being. Psychologists have suggested a few reasons for why we tend to overestimate our future emotional reaction. First, when we predict our reaction to a future event, we focus on very narrow aspects of what life would be like following that event. For example, try to imagine what it would be like if you were bound to a wheelchair. Most of us think of what would change and ignore things that would stay the same. Yes, we would no longer be able to go out for a jog, we would be able to enter only those spaces with wheelchair access, and we would be less independent. However, many aspects of life that give us daily pleasure would remain unchanged. We would still be able to read books, watch movies, go out for dinner, and spend time with friends and family. At first, the changes in our daily life would seem most salient. After a while, however, we would get used to the changes, and the things that made us happy before the incident would again be the ones to take center stage in determining our well-being. Ignoring the elements that would remain unchanged and focusing only on those that would change results in a mismatch between our predictions of how we would feel and how we actually end up feeling.

Not only do we fail to take into account things that stay the same; we also fail to appreciate our remarkable ability to adapt to new circumstances. The human brain is an extremely flexible and adaptive piece of machinery. Think about the last time you went to a movie on an opening night. There were almost no seats left and you ended up in the front row, stretching your neck in an attempt to capture the full screen. At first, it seemed like you would be unable to enjoy the film, or even to see it properly. Within minutes, however, your brain got used to the new form of input; you became engrossed in the film and forgot how unlucky you were to get one of the last seats in the front row.

The mind does better than simply adjust itself to new situations. In order to fully adapt, it creates new capabilities to compensate for those that have been lost. For example, people who lose their eyesight often develop better hearing and a more sensitive tactile sense. Matt lost his physical strength, but he developed his writing ability. When people find themselves alone after the dissolution of a long-term relationship, they quickly develop new skills that previously seemed unnecessary. In any couple, there is usually one partner who is better at cooking, the other more prone to organize social activities or pay the bills. Thus, while there is no need to develop skills that your partner excels in, once that person is gone, you quickly need to figure out how to make an omelette and/or organize your social schedule. As you recognize your newly acquired abilities, you appreciate the positive consequences of the adverse event.

What is remarkable is that reevaluation of adverse events can take place even before the event has occurred. If we know we are about to get fired or be abandoned by a partner, we tend to reframe the incident in our mind in a positive manner in advance. For example, following the economic crisis of 2008, and especially after the fall of Lehman Brothers in September of that year, it became clear that an unthinkable number of individuals were about to lose their jobs. The domino effect did not occur all at once. Most people had some time to consider the upcoming events. Many people who believed they were about to become redundant began to perceive the situation as an opportunity for change and professional growth, rather than as a disaster. They took the opportunity to go back to school, or to look for a better position. Such reevaluation modifies a negative emotional reaction to unemployment before one comes face-to-face with it.[6] This enhances resilience and reduces anxiety. Eventually, if you are laid off, you are better prepared.

How does the brain do this? How do we take unwanted situations and turn them around in our mind? Trying to answer such a question by using experimental methods is somewhat tricky. What my colleagues and I wanted to do was to examine how neural patterns differ when people think about adverse consequences before and after they become an integral part of their lives. It would be somewhat unethical to induce cancer in volunteers or tell them they have been laid off. What to do, then? Our answer? Use one of the greatest human abilities—imagination. Could we trigger real changes in valuation by having people imagine unwanted events occurring to them?

Imagination is a very powerful tool. It is physically impossible for us to experience all of life's possibilities firsthand in order to learn what is good and what should be avoided. Some of life's lessons can be learned from the experience of others, but that is not sufficient for predicting the outcome of every possible situation. To solve this problem, the brain developed a nifty little trick—imagination. Imagination serves an important function by allowing us to simulate and predict the outcome of an infinite number of possible future scenarios. We do this all the time automatically, without even noticing. Before accepting a new job, we imagine what it would be like to work in the new environment, interacting daily with our new coworkers and manager. We tend to quickly simulate almost any action in our mind before acting upon it in real life. From going shopping to bungee jumping, we first run through the event in our mind's eye. Such flexibility goes beyond what can be learned from behavior alone, and it allows us to prepare for what is yet to come.

Humans have become so good at using imagination that we are able to create incredibly genuine images in our mind. These images seem so real that simulating future events enables us to feel the pleasures and pain those events are likely to engender. For example, imagine a large group of ants walking up your thigh. Most of us experience revulsion just at the mere thought. Imagine losing your eyesight. Feelings of fear and sadness are triggered. Although most of us have not experienced permanent loss of eyesight in the past, and may not know someone who has, we can still imagine the situation pretty easily.

Visualizing such adverse events is exactly what we asked our volunteers to do while we recorded their brain activity in an fMRI scanner.[7] We presented them with a variety of medical conditions, such as skin cancer or a broken leg, and asked them to imagine having those conditions next year and tell us how they expected to feel. To determine how miserable they would be with a cast or while undergoing chemotherapy treatment, our participants focused on the negative aspects of the unwanted events. After they had imagined about eighty different dismal events, we introduced an unexpected twist. We presented the volunteers with pairs of conditions they had rated the same and asked them to choose which they would rather have. "If you had to have one of these medical conditions next year—hypothetically, of course—which one would you rather have: A migraine or asthma? A broken leg or a broken arm?" There were forty such pairs. Then we scanned their brains again while they imagined once more having these medical conditions.

Would our subjects perceive the severity of the conditions differently after they chose between them? They did. Although the choices were purely hypothetical, within minutes of choosing the lesser of two evils, the participants' perception of the medical conditions was altered. After selecting one adverse event over the other (let's say Stewart picked fleas over herpes), a participant would rate the selected condition (fleas) as less severe than he had before and the rejected option (herpes) as worse. Although both options may have seemed severe at first glance, by reevaluating the condition he selected and viewing it in a positive light ("Fleas are not that bad; I will simply buy hydrocortisone cream for the itching and call a professional exterminator"), Stewart experienced a heightened sense of well-being.

The transformation in how Stewart and others viewed fleas, herpes, and other misfortunes was also apparent when examining their brain activity. Before Stewart was presented with the demanding task of choosing between fleas and herpes, there were no detectable differences in the pattern of his brain activity while he imagined having fleas or herpes. After selecting one uncomfortable condition over the other, though, changes suddenly emerged. The alterations we observed were similar to the ones mentioned in the chapter where subjects had to pick one vacation destination over another. There, we saw an enhancement in caudate nucleus activity when individuals fantasized about the selected holiday destination as opposed to the rejected option. To refresh your memory, the caudate is a nucleus deep in the brain that is involved in signaling expectations of emotional outcome. In this study, after backaches were selected, let's say, rather than migraines, enhanced caudate activity was detected while participants imagined having backaches relative to migraines. The caudate, most likely, was updating the new value associated with backaches—from "ouch . . . terrible" to "not great, but not too bad."

There was one other change that was revealed by the fMRI data. It was in the same region we had previously identified as key in mediating optimism—the rostral anterior cingulate cortex (rACC).

In the "Pick an Illness" study, after Annabelle, for example, decided that having gallstones was better than having kidney stones, enhanced activity was detected in her rACC when she imagined having gallstones. The rACC is thought to track the salience of stimuli by monitoring signals from brain areas that process emotional and motivational information and to modulate these responses.

Imagine you find yourself in a wheelchair. At first, negative emotional reactions (such as panic and fear) and negative thoughts ("I will never be able to run on the beach again") arise. In order for you to cope, however, these thoughts are inhibited and attention shifts to the positive ("At least I have a loving family and a clear mind"). The rACC is key in this process.

Extinguishing Fear

The neural mechanisms I have just described are very similar to the ones responsible for extinguishing fear. In order to study how fear arises, scientists often use a simple experimental paradigm known as *fear conditioning.* In this paradigm, a person (or in some cases a rat) is given a small electric shock (or some other aversive treatment) following the presentation of an otherwise neutral stimulus such as a sound. An experimenter may introduce a high tone and then give the participant a small electric shock. The participant rapidly learns that the presence of a high tone means that an electric shock is coming. Eventually, the high tone alone is sufficient to produce fear responses, such as an elevated heart rate and sweating.

In real life, tones and shocks don't often go together, but you can imagine many instances when two stimuli become associated. For example, as a child I would walk to school every morning. To get there, I had to pass through a relatively quiet street. The first few months at my new school, the journey was uneventful. Then, one day, I encountered a terrifying-looking dog that barked in rage, flashing his sharp teeth and drooling saliva. Given the fact that I was a small kid, the dog seemed enormous. I was petrified. Exactly the same chain of events occurred on several consecutive days. Eventually, just upon entering the street I would start sweating and my heart rate would rise, regardless of whether the dog was even in sight.

A few weeks later, the dog disappeared. Maybe his owners decided to keep him indoors, or maybe he was attacked by an even larger, scarier-looking animal. I don't know. When a couple of weeks had passed without any sign of the dog, I no longer thought twice about walking down that street. My breathing didn't become heavy and my pulse did not rise the minute I entered the once intimidating street. This is what psychologists call fear extinction—learning that a stimulus that once predicted an adverse outcome no longer does.

In a series of experiments, Elizabeth Phelps and Joe LeDoux of New York University showed that the rACC, as well as other parts of the ventral medial prefrontal cortex, were responsible for inhibiting the fear response that was generated by the amygdala.[8] The amygdala is critical for producing fear reactions to a conditioned stimulus (such as the tone or, in my case, the street) due to learned associations (tone equals shock, or street equals dog). When the stimulus is no longer a valid sign of danger, the fear response will be switched off, and the rACC is key in that process. If we did not have a mechanism that extinguishes fear when it becomes irrelevant, we would probably be walking around terrified twenty-four hours a day.

We don't. Most of us walk gaily down the street, and when we encounter events that trigger anxiety, we often make a conscious effort to calm ourselves down. Imagine yourself on a plane that suddenly starts shaking. Your amygdala jumps into action and you begin sweating, but then you tell yourself, This is just turbulence. I have experienced it many times before and survived. I'll just lie back and enjoy the ride. What you are doing is regulating your emotions. Your ventromedial prefrontal cortex (VMPFC) is inhibiting the response of the amygdala and your heart rate becomes stable. There are many ways we can control our emotions: We can repress certain thoughts and generate others (such as thinking of a happy incident to avoid crying in public following a disappointing encounter), focus on certain aspects of a situation or stimulus, or positively reframe events as Matt did when he found himself in a wheelchair.

Biased Perception

Regulating our emotions, reframing events, and inhibiting fear change the way we think about the world. Will doing so also change the way we visually perceive the world? Should the phrase "viewing the world through rose-colored glasses" be taken literally?

Consider the mirage illusion—that is, the false sense of seeing water in the desert. An article published in 2001 reported that you do not need to be in the desert for thirst to alter your perception.[9] People who are thirsty are more likely to perceive transparency, a characteristic associated with water, in an ambiguous visual stimulus than people who are hydrated. Presumably, the overwhelming desire for water influenced the way participants saw their environment, perceiving fluids when there were none. Take another example: our tendency to mistakenly see a loved one in the crowd—particularly in instances when we wish for such an encounter. Our desires influence our visual perception in a manner that fits our goals, tricking us into seeing water when we are thirsty or our lover when we are lonely.

In a creative, and somewhat outrageous, experiment, Emily Balcetis and David Dunning showed that perception can be altered by wishing.[10] They asked students to dress up as Carmen Miranda, a Brazilian singer and actress in the 1940s and 1950s who was well-known for wearing a Brazilian-inspired costume that included a hat with various tropical fruits on top. She would wear this distinctive hat in movies and Broadway shows, mostly musicals. The students were to walk across the quad on campus (365 feet in each direction) wearing a grass skirt, a coconut bra, and a plastic-fruit hat. The scenario could easily have been taken out of a familiar nightmare. You know the one: You are standing naked in front of your peers, while they are fully dressed, staring at you in disbelief. Granted, the participants were not naked; they were wearing coconut bras, grass skirts, and fruit hats to cover their flesh. Sounds just as bad to me. Amazingly, the researchers were able to find thirty-two students who volunteered (yes, volunteered) to participate in the crazy scheme in exchange for . . . no, not a few thousand dollars, but merely course credit.

Before sending the students off to embarrass themselves in public, the researchers told half of the participants that they could choose to perform another task instead if they so desired. They then signed a waiver entitled "Freedom of Choice." Those participants were dubbed the "high-choice group"; they had chosen the task themselves. The other participants were told that the researcher had picked the task for them from several options.

They signed a waiver entitled "The Experimenter Choice"; they were the "low-choice group."

Both groups were sent off to trot up and down the quad in their new attire. They were then asked a simple question: "How far was the distance you walked?" As I mentioned, the quad was 365 feet long. After walking that length in coconut bras and plastic-fruit hats, all participants perceived what their hearts desired: They significantly underestimated the length of the quad. The students in the high-choice group underestimated the length the most, guessing, on average, a distance of only 111 feet. Participants in the low-choice group estimated the distance as 182 feet.

The researchers were not quite done. The Carmen Miranda experiment went so well, they decided to come up with yet another fun task for the students. This time, the students were permitted to remain in their own clothes. They were asked to kneel on an all-terrain skateboard and push themselves up a hill with their hands. Again, they were assigned to either a high-choice group or a low-choice group. They were then asked to estimate the slope of the hill they were about to climb. Once more, participants who had chosen the task themselves perceived the task as less aversive, estimating the slope at twenty-four degrees, while participants who were assigned the task by the experimenter estimated the slope to be thirty-one degrees.

The conclusion from both the "Walk Around in a Coconut Bra and Fruit Hat" study and the "Push Yourself Up the Hill with Your Hands" study was that our aspiration to achieve positive outcomes and avoid negative ones is so robust that it alters the way we visually perceive our surroundings. That's not all. People are more likely to perceive their environment inaccurately, finding it less intimidating (such as estimating the distance to be traveled as shorter, and a hill to be climbed as shallower) if they had selected the unpleasant task themselves.

The Role of Cognitive Dissonance

Why does selecting an unpleasant task make it more pleasant? Choosing to complete an embarrassing or physically demanding assignment conflicts with prior belief ("Walking around campus half naked with a fruit hat is to be avoided"). Such conflict, also known as cognitive dissonance, produces a negative arousal. This uncomfortable emotion can be reduced by biasing our perception of the environment ("Well, the distance to be walked is quite short, I can do it quickly, and no one will notice").

In the 1950s, there was a cult in the United States that believed the Earth would be destroyed by aliens on December 21, 1954. In his book *When Prophecy Fails,* Leon Festinger, the father of cognitive dissonance theory, told the story of this cult.[11] According to the cult's belief, the aliens had sent the doomsday information to the leader of the group—a Mrs. Keech—with a promise that members of the cult, and they alone, would be spared. Alas, when December 21 came and went with no aliens in sight, it became obvious that the end of the world was not so near after all. However, instead of abandoning the cult, the members became increasingly committed to it. The notion among the members was that their unshaken faith had saved the world from destruction and all was well yet again.

The members' strong belief in their leader and her prediction conflicted with reality (the Earth was still there). Such conflict triggered dissonance, which was resolved by adopting a new belief: "The leader was right, and because of our belief in her, the world was spared."

For cognitive dissonance to have been triggered in the first place, it was critical that the members had freely chosen to become part of the cult and had no ulterior motive to join it. If they had been forced into the cult by other members, or if they had been promised money for joining, the need for dissonance reduction would never have emerged. This is because their membership could then have been explained by other motives (money, force) rather than by faith alone and thus would not have conflicted with the reality that the world had not come to an end.

Festinger went on to demonstrate his theory in the lab. Rather than convincing people that the end of the world was near, he simply asked participants to perform a boring task for an hour. They had to turn pages of a book—not an enjoyable task by any standard. The participants then had to persuade the next volunteer that the task was, in fact, pleasant. They received either one dollar to persuade the other person of the desirability of the task or twenty dollars to do so. Then they were asked to rate how engaging the task really was.

Participants who had received only one dollar rated the "Turn the Page" task as more interesting than participants who had received twenty dollars. What was going on? Why did the volunteers with *less* money in their pockets walk away believing that the last hour had, in fact, been enjoyable? It seems that the deprived participants, those who got only one dollar, had dissonance to resolve. Here was the source of conflict: On the one hand, the participants believed the task was boring; on the other hand, they had persuaded a fellow student that the task was engaging—and apparently for no good reason at all! That must mean that the task was not as bad as they had originally thought. Those who had received the larger amount of money, however, easily resolved the conflict by explaining their deceptive actions as a necessity. They'd had to persuade the other student that turning pages was fascinating in order to receive the large reward (twenty dollars). Those who had received only one dollar had had to find another justification for their actions, and so they'd simply changed their valuation of the task from "boring" to "not so bad." Conveniently, inconsistency was reduced.

So once again it is apparent that the human mind finds a quick and easy way to restore balance. If we change our attitude, we regain well-being. In physics, the principle of relativity requires that all equations describing the laws of physics have the same form regardless of reference and frames. The formulas should appear identical to any two observers and to the same observer in a different time and space. Attitudes and values, however, are subjective to begin with, and therefore they are easily altered to fit our ever-changing circumstances and goals. Thus, the same task can be viewed as boring one moment and engaging the next. Divorce, unemployment, and cancer can seem devastating to one person but be perceived as an opportunity for growth by another person, depending on whether or not the person is married, employed, and healthy. It is not only beliefs, attitudes, and values that are subjective. Our brains comfortably change our perceptions of the physical world to suit our needs. We will never see the same event and stimuli in exactly the same way at different times. Two minds may perceive the same hill, but one will view it as steep and the other as shallow, according to what the person expects to find at the top. A quad can seem long to one person and short to the next, depending on whether or not the person chooses to stride across it wearing a coconut bra, a grass skirt, and a plastic-fruit hat.

CHAPTER II

A Dark Side to Optimism?

From World War II to the Credit Crunch—Underestimating Risk Is Like Drinking Red Wine

Leopold Trepper was a Soviet spy. It was the beginning of World War II and Trepper was situated in Brussels, posing as a Canadian industrialist. His cover was an export business called the Foreign Excellent Raincoat Company, which had branches all over Europe. Trepper, however, was providing his employers with more than waterproof garments. He would present the Russians with top-secret data that could change the course of history—information that could alter one of the most lethal battles known to man. Unfortunately, the information Trepper had to offer was troubling. In fact, it was so dire that it was ignored and brushed away in disbelief—an oversight that may have cost the Soviet Union countless military and civilian casualties.[1]

Trepper's code name was "Leiba Domb," but even his birth name did not reveal his origins. He held a proper German surname and a given name no different from that of the Holy Roman Emperor—Leopold I—the king of Hungary in 1655. Trepper, though, was far from a Christian emperor. He was a Jew born in a poor Austro-Hungarian town called Nowy Targ. When he was a child, his family moved to Vienna, where it soon became clear that Trepper was not one to sit quietly while the world around him was shaking. He had a political soul. He first joined the Bolsheviks, members of the Communist Party, and at age nineteen was sent to prison in Poland for organizing a strike. When released, he headed to Palestine, where he joined the Hashomer Hatzair, a Zionist socialist movement that fought against the British Mandate of Palestine. After being expelled from Palestine, he moved to France, only to escape yet again when his political organization was uncovered by French intelligence.

That was when Trepper found himself in Moscow, where he quickly became an agent for the general staff of the Russian armed forces. His task was to manage and direct a Soviet intelligence group in Nazi-occupied Europe; this group was known as the Red Orchestra.

While Trepper was undercover as a businessman in Europe, the Germans were preparing the largest military offensive in history—an invasion of the Soviet Union. Hitler had declared his intention to invade the Soviet Union back in 1925 in his book *Mein Kampf,* and fifteen years later he was ready to turn his visions of attack and conquest into reality. In December 1940, he received and approved Directive Number 21, the plans for what is now known as Operation Barbarossa.[2]

Trepper promptly warned the Russians of Hitler's objectives:

> In February, I sent a detailed dispatch giving the exact number of divisions withdrawn from France and Belgium, and sent to the east. In May, through the Soviet military attaché in Vichy, General Susloparov, I sent the proposed plan of attack, and indicated the original date, May 15, then the revised date, and the final date.[3]

The final date was June 22, 1941. The Soviet Union was invaded by 4.5 million troops, leaving no question as to the intentions of the Nazi leader. Back in February, though, Stalin believed that the Germans would not attack.[4] The two countries had relatively strong diplomatic and economic relations, as well as a signed agreement. The Molotov-Ribbentrop Pact, a nonaggression treaty, had been signed by the Soviet Union and Germany in 1939. It secretly outlined the division of border states between them. While Stalin might not have perceived the Germans as the most trusted of allies, he could not imagine they would stab him in the back, either. In fact, he was so enraged by Trepper's suggestions that he instructed that the spy be punished for his lies.

If Trepper had been the only bearer of unwanted news, one could understand Stalin's reluctance to listen. What weight did one spy carry relative to a signed agreement between two nations? But Trepper was not alone. Shortly after Trepper first warned Moscow of the planned attack, another Soviet spy, Richard Sorge, also known as "Ramsay," informed Stalin's government that 150 German divisions were gathered along the border. Sorge eventually provided Stalin with the exact date of the Nazi invasion, but he, too, was ignored. So was President Roosevelt, who gave the Russian ambassador data collected by American spies regarding the operation.[5] Stalin turned a blind eye to the cruel reality. "He who closes his eyes sees nothing, even in the full light of day. This was the case with Stalin and his entourage," Trepper wrote.[6]

While in many respects Stalin and his staff were outliers, exhibiting terrifying behaviors and beliefs—ones that are, we hope, rare—at least in one way their minds worked in a very predictable, very common manner.

Burying Your Head in the Sand

Consider the following short list of events. Try to estimate your probability of encountering these events in your lifetime (if you have already experienced some of them, assess your likelihood of experiencing them again). How likely are you to:

1. have cancer?
2. get a divorce?
3. lose your job?

Let's consider the first question. What probability did you assign it? In the United States, cancer, in its different forms, accounts for approximately a quarter of all deaths.[7] The likelihood of having cancer in your lifetime is, of course, larger—roughly 33 percent. Was your estimate higher or lower?

Just as the Russians underestimated the likelihood of a German invasion, most of us tend to underestimate the probability of negative outcomes in our lives.[8] For the first question (the likelihood of having cancer), most of us would assign a probability lower than 33 percent, and for the second question (the likelihood of getting a divorce), most of us would assign a probability lower than 50 percent (yes, in Western culture, about 50 percent of all marriages end in divorce).

In a series of studies, Neil Weinstein (who coined the term *optimism bias*) showed that people believe they are less likely than average to suffer misfortunes (such as being fired from a job, being diagnosed with lung cancer, developing a drinking problem). Simple math will show that if *most* people claim their chances of experiencing a negative life event are *less* than average, then clearly they are wrong. We can't *all* do better than the average Joe.

Maybe not, but deep down we believe we can. We truly think our children will grow up to be healthy and successful. And when standing at the altar or civil registry desk, we expect to be blissfully married for the rest of our lives. And yet, half of us are wrong. The decree nisi is so common that, in the words of Oscar Wilde, "the world has grown suspicious of anything that looks like a happily married life."

Some of you may not find this particular fact very shocking.

Personal experience has taught us that falling in love leaves little room for statistical calculation, or, indeed, for rational thought of any kind. And even the experience of breakup or divorce won't necessarily darken our bright outlooks too drastically—at least not enough to prevent us from trying again. The high rates of remarriage in the population suggest that even after we crash and burn once, twice, thrice, we still believe the next time will be better. Remarriage, as Samuel Johnson described it, is "the triumph of hope over experience."

Are we just ignorant of the high divorce rates out there? Or is it that we all simply expect *our* relationship will be, well . . . *different*? In 1993, the psychologists Lynn Baker (University of Texas) and Robert Emery (University of Virginia) decided to investigate this matter closely.[9] They sought out individuals who were planning to get married and asked them to estimate the divorce rates in the United States. As it turned out, most people were pretty good at gauging the overall likelihood of divorce. The academics then asked their subjects about the expectations they had for their *own* marriages. Almost uniformly, these individuals were idealistic about the longevity of the union ahead of them. Not only did they underestimate the likelihood of their marriage ending in divorce; they also underestimated the range of negative consequences they would encounter should their marriage break up. What if we were to significantly increase people's awareness of the incidence of divorce? Would their rosy outlook be despoiled? The answer, according to Baker and Emery, is no. They found that taking a course on family law did nothing to diminish the unrealistic optimism of law students engaged to be married.

Just as Stalin ignored Trepper's warnings, the law students did not see how the common negative consequences of divorce could have anything to do with their own future. In other words, even when detailed, reliable information is presented, such as the average likelihood of divorce or the exact date of a German invasion, people often look the other way, hanging on to a brighter outlook.[10]

Stalin was not the only one to look away during World War II. On the other side of the border, the German commander was also becoming overconfident, ignoring warnings from his advisers. Hitler expected a rapid triumph in the fight against the Soviet Union.[11] He did not predict the long, lethal battle that lay ahead, and thus he did not prepare for a war that would last into the cold winter months. His lack of adequate planning meant that when summer and fall passed and fighting was conducted in subzero temperatures, his troops were ill prepared. They did not have the necessary clothing and equipment to endure the harsh environment. He miscalculated not only the duration of the operation but also the financial consequences. Although he was cautioned in advance of the huge economic cost Operation Barbarossa might entail,[12] he maintained that "from now onwards he wasn't going to listen to any more of that kind of talk and from now on he was going to stop up his ears in order to get his peace of mind."[13] That peace of mind was to be short-lived.

Unlike the Soviet and German commanders, most of us do not decide the fate of a nation. We do not need to make predictions that will determine whether a country should go to war or whether or not to prepare for an invasion. Nevertheless, as Weinstein and others have shown, when it comes to our relationships, our health, and our careers, we, too, underestimate possible pitfalls. These expectations will determine our choices and alter the course of our lives. For example, if we expect a long, happy marriage, we may fail to sign a prenuptial agreement and thus might find ourselves in the midst of a messy divorce battle. On the other hand, if we do not hold positive expectations about the longevity of our relationship, we might never take the plunge.

How Do We Maintain Optimism in the Face of Reality?

Commonplace as the optimism bias may be (according to data gathered by the Yale psychologist David Armor, about 80 percent of the population holds optimistic life expectations), the phenomenon is baffling. What is puzzling is that we go about our daily business, experiencing both negative and positive events—we read the newspapers; we know that global economy is in trouble, as is the environment; we are aware of the many risks out there, such as cancer and AIDS—and yet we underestimate our likelihood of getting stuck in traffic, experiencing heartache, or being attacked by a war-seeking commander.

According to prominent learning theories, a human—or any animal, in fact—should learn from negative (and positive) outcomes and correct his or her expectations. Why is it that we do not learn?

Think back to the study I conducted at the Weizmann Institute in Israel (described in the prologue). The students at the Weizmann Institute overestimated the probability of positive everyday events happening to them in the upcoming month (such as having a positive sexual encounter or enjoying a party) by about 20 percent. Although they had accumulated years and years of daily life experiences and really ought to have been able to predict more accurately the likelihood of everyday occurrences in the upcoming month, they still were unrealistically optimistic.

To make relatively accurate predictions, all we need to do is look back in time and say, "Last month I was late for most meetings, enjoyed only half the films I viewed, and received no gifts from my loving partner. Therefore, this month I am likely not to be on time for the majority of my appointments, will probably appreciate only 50 percent of the movies I watch, and should not expect any gifts." Generally, accurate processing of information is crucial for optimal behavior. This raises the question of how unrealistic views of the future persist when information counter to those beliefs is abundant and available. My student Christoph Korn and I searched for an answer.

This was our reasoning: If humans maintain an unrealistic view of the future, even when accurate data is available, it must be the case that the brain processes information regarding the future in a selective manner. A learning bias, if you will. A bias that allows us to incorporate desirable information into our future outlook but not undesirable information—resulting in optimism. Is that how the brain works? And if so, why?

Earlier in this chapter, I asked you to estimate the likelihood of experiencing a range of negative events (cancer, divorce, job loss). I conducted a similar exercise with a group of volunteers. I asked them to estimate the likelihood of experiencing eighty different adverse life events (e.g., breaking a limb, missing a flight, being in a car accident) while I recorded their brain activity using an fMRI scanner.

After the volunteers estimated the likelihood of experiencing the adverse events and while they were still being scanned, I provided them with information regarding the average probability of that event occurring to a person in the developed world, just as I did for you earlier in this chapter. Would the participants learn from the feedback provided? Would they update their expectations? These were the critical questions. We wanted to know what people would do with the information presented to them. Does the brain process desirable and undesirable information differently? Could this (at least partially) explain the optimism bias?

After the participants received the statistics regarding the probabilities of different negative events, they were asked again to assess their own likelihood of experiencing those events. In general, participants learned from the feedback provided. However—and this is the crucial finding—differential learning was observed when participants received desirable and undesirable information regarding the future. If Jane estimated her probability of getting an ulcer at 25 percent and then learned that the average probability was only 13 percent, she would be more likely to update her estimate the second time around (maybe estimating her likelihood of getting an ulcer at 15 percent). If, however, she initially estimated her probability of getting an ulcer at only 5 percent and learned that the average probability was much worse—13 percent—she would adjust her estimate only a small amount, if at all.

Could differential memory of desirable and undesirable information explain this selective update of expectations? In other words, did Jane remember positive statistics but not retain information that challenged her optimistic outlook? That was not the case. Participants remembered the probabilities presented to them equally well whether the information was desirable or not. As in the study conducted by Baker and Emery, the participants had no problem reporting the accurate average probabilities of the unwanted events. It was not the processing of the information per se that was biased, but the utilization of that information. When the data was better than expected, people took notice and incorporated it into their outlook ("Ah, the likelihood of death before sixty is only 10 percent. . . . I may live longer than I expected"); when it was worse, it was discarded ("Mmm . . . 23 percent chance of having a stroke. That cannot be relevant to me—I am in excellent health"). Exactly the same information was perceived as significant or irrelevant, according to whether it was better or worse than expected.

The participants' brain activity gave us clues to exactly what was going on. Normally, when we have a certain expectation, the brain tracks the difference between that expectation and the outcome.[14] Let's say you are having dinner at a new restaurant and the waiter describes the daily special—lobster ravioli. The price of the dish is not listed. So while the waiter is talking, you are busy calculating the cost of this delicacy—twenty-seven dollars, you guess. The waiter finishes describing the creamy sauce and declares the price—thirty-five dollars. You are a bit surprised, and the mismatch between your prediction and the outcome is represented by an enhancement of brain activity. The larger the discrepancy, the larger the brain signal. This "mismatch signal" is crucial. It is used by the brain to learn—to update expectations. Next time you show up at the restaurant, not only will you know the exact cost of the creamy lobster dish but you will now have a benchmark. When a new dish is offered—oyster fettuccine—you will probably do a better job at estimating the price.

Similarly, when we presented our volunteers with statistics regarding the prevalence of the negative events, we observed activity in their frontal lobes that was tracking the difference between their estimates and the statistics provided. So when Howard, one of the participants, estimated his likelihood of having genital warts at 20 percent and learned that the average probability is lower—about 12 percent—an enhanced BOLD (blood-oxygen-level-dependent) signal was observed in parts of his frontal lobe. If the mismatch was greater (let's say he had originally given an estimate of 30 percent), the signal would be even greater.

We already knew that the brain tracks errors in prediction, so this was not surprising. What was unexpected was that the brain was pretty good at tracking mismatches *only* when the new information was positive (such as in the example above). When the news was undesirable—when a person estimated a 1 percent chance of suffering from genital warts and learned that the average probability is about 12 percent—the brain did not track the error as closely. Since the frontal lobes selectively recorded desirable errors but failed to record undesirable ones, people learned more from the good news than from the bad news. As a result, the participants left our lab even more optimistic than when they'd arrived!

The Advantage

Is this a good thing? If we underestimate health risks, our likelihood of seeking preventive care and medical screening is reduced, and the likelihood of engaging in risky behavior is increased.[15] How many times have you not bothered to apply sunscreen on a hot day, telling yourself you're unlikely to get skin cancer from a single exposure? What about skipping a scheduled medical checkup, assuming that everything is fine? Or practicing unsafe sex? Underestimating risk can lead to an infinite number of medical issues that otherwise could have been prevented, costing our health system millions of dollars a year.

Why would our brains be wired in a way that biases the process by which we learn about the world around us? Why would we develop a system that causes us to predict the future inaccurately? Could being irrationally optimistic have survival value?

As I outlined in chapter 3, optimism can be a self-fulfilling prophecy. Take, for example, a study that tracked 238 cancer patients. Astonishingly, the study revealed that pessimistic patients under the age of sixty were more likely to die within seven months than optimistic patients whose health, status, and age were initially the same.[16] Optimists also experienced faster recovery after coronary bypass surgery than pessimists, and were less likely to be rehospitalized than pessimists.[17] There may just be a reason for the brain's failure to record undesirable information regarding the future. Underestimating the probability of future adverse events reduces our level of stress and anxiety, which is beneficial to our health.

There are other advantages to being optimistic that most of us would not have expected. Take a look at the following list and try to guess which of the factors can be predicted by a person's level of optimism.

1. Number of hours worked per day
2. Having a savings account
3. Liking ice cream
4. Marital status
5. Expected age at retirement
6. Smoking habits
7. Strong bond with one's laptop

For each of the factors you believe is related to optimism, try to predict the direction of the correlation. Are optimists more likely to smoke, or less? Do they expect to retire later in life, or earlier? Do they like ice cream, or dislike it?

Manju Puri and David Robinson, economists at Duke University, wanted to study the relationship between optimism and the choices we make in life.[18] For their investigation, they used data from the U.S. Federal Reserve Board's Survey of Consumer Finances. This survey includes many questions related to people's work habits, spending and saving behavior, health-related behavior, and future expectations. To measure optimism levels, Puri and Robinson focused on responses to the question "How long do you think you will live?" People usually overestimate their longevity by a few years. By looking at the difference between people's self-reported life expectancy and their actual life expectancy as presented in actuarial tables, the academics had a good indicator of people's optimism. They knew misjudgment of life expectancy was a good measure of trait optimism because it has been shown to correlate with the standard psychological optimism tests.

They divided respondents into extreme optimists, moderate optimists, and pessimists. People who overestimated their life expectancy by about twenty years (around 5 percent of those surveyed) were labeled "extreme optimists." The "moderates" accounted for the vast majority. They were the ones who overestimated their life expectancy by only a few years. People who underestimated their longevity were labeled "pessimists"—they were a minority.

Let's look back at the list of factors that may be predicted by optimism. Optimism was found to be related to numbers 1, 2, 5, and 6: number of hours worked per day, having a savings account, expected age at retirement, and smoking habits. Moderate optimists worked longer hours, expected to retire later in life, saved more (with longer planning horizons), and smoked less than all other individuals. Extreme optimists worked fewer hours, saved less, and smoked more.

From investment choices to productivity, optimism turned out to be a crucial factor. Moderate optimism correlated with sensible decisions, while extreme optimism correlated with seemingly irrational decisions. As in almost everything in life, moderation seemed to be key.

A certain underestimation of the hurdles in front of us allows us to jump forward with force. If, however, we ignore risks and hazards altogether, assuming they are irrelevant for us, we will be ill prepared when hurdles materialize. The conclusion, as Puri and Robinson aptly put it, is that "optimism is like red wine: A glass a day is good for you, but a bottle a day can be hazardous." Extreme optimism, like excessive drinking, can be dangerous not only for our health but for our pockets, too.

The Pitfall

Consider the Sydney Opera House.[19] On September 13, 1955, the premier of New South Wales, in Australia, Joseph Cahill, announced a competition for the design of a new opera house to be situated on Sydney's Bennelong Point. More than 230 architects from around the world submitted proposals. Out of these, Jørn Utzon, a Danish architect, and his team were selected. They were awarded an opportunity to build what would later be referred to as a modern masterpiece, an iconic building of our time. Utzon and his colleagues went to work straightaway. Their estimated cost for the project was seven million dollars, and the planned completion date was January 26, 1963.[20] This would give the team approximately six years to complete the project. At the time, six years seemed a reasonable period in which to build the opera house. Soon enough, however, unexpected hurdles materialized.

First, the team faced unforeseen stormy weather and had difficulties with redirecting storm water away from the building site. In addition, they were forced to begin construction before the final drawings had been completed. This led to multiple problems, including the creation of weak podium columns, which could not support the roof and had to be rebuilt.[21] By 1966, the project was already running sixteen million dollars over budget, and the team was more than three years behind schedule. The tension between the architects and government officials was rising daily. Each party was blaming the other for the dismal situation. Finally, Jørn Utzon resigned, and the completion was delayed once more.

In 1973, a decade after the original due date, the Sydney Opera

House was completed. The cost was $102 million—more than fourteen times the original budget! No doubt the end product is impressive, but could there have been a better plan? Could possible difficulties have been foreseen and accounted for? Could a more reasonable budget and time line have been proposed?

The Sydney Opera House is not an isolated case. Whether it is a construction job, a film, a theater project, a dinner party, home renovation, war, or a peace plan, cost overruns and delays in implementation are the norm. The British government, for one, has decided to address this problem. Specific guidelines on how to correct for the optimism bias in appraisals were published in the British government's Green Book, which provides an overall methodology for economic assessment. A special supplementary guideline on the issue states: "There is a demonstrated, systematic, tendency for project appraisers to be overly optimistic. To redress this tendency appraisers should make explicit, empirically based adjustments to the estimates of a project's costs, benefits, and duration."[22] Adjustments for the optimism bias have since been factored into the budget of many of the government's projects, including, most recently, the 2012 London Olympics.

Suppliers of credit lines seem to be aware of the optimism bias (although not necessarily when it applies to them). Dare we say that they seek to utilize and enhance said bias in the marketing of credit products? Attributing unrealistic low probability to negative life events (such as illness and job loss) and unrealistically high probability to positive life events (such as getting a raise) triggers debtors to borrow more than they would have borrowed otherwise. And yet, as was bluntly apparent in September 2008, stocks stubbornly persist in going down as well as up. Economists have suggested that the optimism bias was a root cause for the financial meltdown of 2008.[23] The optimism bias was not only blurring the vision of the private sector (individuals who believed that the value of their homes and salaries would grow but that interest rates would remain constant) but also that of government officials, rating agencies, and financial analysts, who continually expected improbably high profits.

You may think that in light of 2008's credit crunch and the generally pessimistic outlook portrayed in the media at the time, people would have had bleak expectations of the financial future of their businesses. This was not the case. According to a survey conducted in July 2008, 76 percent of the 776 U.K. business professionals surveyed still felt optimistic about the next one to five years. Although the respondents were well aware of the bleak economic climate at the time, when they looked to the future, they did not see poverty and bankruptcy for themselves. Why?

When people imagine adverse situations, they see themselves getting out of the muddle. Although the businessmen surveyed may have been experiencing losses, when they closed their eyes, they envisioned ways in which they could rebuild their businesses and eventually profit. Consider the response of a volunteer in a study I conducted. When asked to describe what it would be like for her to lose her apartment keys, she said, "Getting locked out is always very annoying, but I always have a spare key somewhere and/or [one] with someone (a roommate). Even though I have never owned an apartment, I assume the landlord has a key, so I envisioned myself going downstairs and asking for a spare."

I have been locked out of my apartment more times than I care to remember. Getting back in was never as simple as this respondent imagined it to be. Once, my dear sibling had to drive an hour and a half with the spare key. Another time, I was locked out of my apartment in central London while taking out the trash. I had to knock on my neighbor's door (this was someone whom I had never met, as I had just moved in a few weeks earlier) and ask to use the phone. Luckily, my neighbor turned out to be a lovely lady. An hour later, the locksmith arrived. I was back in my apartment—alas, a couple of hundred pounds short.

Clearly, imagining how things may go wrong (being locked out of your apartment) can aid in identifying actions (giving a spare key to your neighbor) that will help us avoid adverse situations; it may also aid in preparing us emotionally for disappointment and heartache. However, *pondering* unpleasant events interferes with daily activities by promoting negative effects such as anxiety and depression.

"For myself I am an optimist—it does not seem to be much use being anything else," said Winston Churchill to the gathering at the Lord Mayor of London's banquet in 1954. A pessimist, in Churchillian terms, will see the difficulty in every opportunity and thus will be unlikely even to try, while an optimist will see the opportunity in every difficulty.

Yes, the 2012 London Olympics budget had to be adjusted to account for overoptimistic predictions. But if the human spirit were not an optimistic one, would there be anyone around to participate in the actual games? My guess is that the number of athletes who expect to win a medal at the Olympics significantly exceeds the number of contestants who will mount the podium to be garlanded in due course. Most athletes subject themselves to years of intensive training because they can clearly envisage the end goal—and it is a magnificent one.

EPILOGUE

A Beautiful Mademoiselle or a Sad Old Lady?

From Prediction to Perception to Action

We have journeyed from the dark skies of Sharm el-Sheikh to the crowded lockers of the Los Angeles Lakers, from an Irish pub to the campus of UC Davis and into a black London cab. This path was taken in order to make two major claims.

The first argument this book makes is relatively simple: Most of us are optimistic. Although good things may transpire, on *average* our expectations exceed future outcomes.[1] We are not necessarily aware of our bias. Like other illusions of the human brain, the optimism bias is not easily accessible for introspection.[2] Nevertheless, science has shown us that our minds tend to engage in thoughts of sunny days.[3] We think of how well our kids will do in life, how we will obtain that sought-after job or that house on the hill, how we will find perfect love and happiness. We imagine our team winning the critical game and look forward to a relaxing vacation in Costa Rica. We visualize our investments yielding nice profits and the value of our homes growing. Even when financial markets crash and war-seeking commanders threaten to take over, our instincts tell us we will endure.

Don't get me wrong: Our minds do entertain darker thoughts. We worry about losing loved ones, failing at our job, or dying in a terrible plane crash over the Red Sea. Research shows, however, that most of us spend less time mulling over negative outcomes than we do over positive ones, and when we do contemplate defeat and heartache, we tend to consider how those can be avoided.

Although we are optimistic, our expectations do not usually border on lunacy. Most of us do not anticipate winning an Olympic gold medal, being the president of the United States, or becoming a Hollywood star. An optimism bias simply means that, more often than not, our expectations are slightly better than what the future holds. Overall, this is beneficial. Data pointing toward the upside of optimism is plentiful; optimists live longer, are healthier and happier, make better financial plans, and are more successful.[4]

This brings us to the second claim of the book—the assertion that our brains have evolved to overpredict future happiness and success, because, funnily enough, doing so makes health and progress more likely. Understanding how the mind generates and maintains unrealistic optimism and how—even more bafflingly—optimism leads to professional and personal success requires intimate knowledge of how the human brain works. The tendency for positive predictions to create positive outcomes (whether subjective or objective) is rooted in fundamental rules governing the way the mind perceives, interprets, and alters the world it encounters.

The brain is organized in a hierarchical structure. It is this precise arrangement that allows our expectations to influence both our *perception* of reality and our *actions*—thereby altering reality itself. In this book, we have focused on structures at the apex of the brain's hierarchy, such as the frontal cortex, and on evolutionarily older structures that are found lower down the ladder. As you might remember, the frontal cortex carries out higher cognitive functions, such as planning, abstract thinking, theory of mind (thinking about what others are thinking), error detection, and conflict resolution.[5] Moving deeper into the brain, we find the subcortical regions. These include structures we have considered repeatedly throughout the book, such as the amygdala, which is involved in emotional processing,[6] the hippocampus, which plays an important role in memory,[7] and the striatum, which is key in representing the value of stimuli and actions.[8]

Via neuronal signaling, higher layers of the brain can convey expectations to lower levels, biasing their activity.[9] Earlier in the book, I asked you to close your eyes and imagine your future. I knew, from research I had conducted, that you were more likely to imagine good things than bad. Brain-imaging data suggests that this imbalance is due to neurons in the frontal lobe altering the activity of subcortical regions, enhancing signals that convey positive emotions and associations while you consider the future.[10] In a feedback loop, neurons from lower structures project information back to higher structures, strengthening and confirming initial expectations.

Let's use a visual example (as we did in chapter 1) to illustrate how our expectations alter the way we perceive and interpret the world. Take a look at the beautiful young girl dressed in feathers, who is portrayed in the figure below.

Figure 3

Adapted from *Puck* magazine (1915).

Can you see her? Great. Now look at the figure once more. But this time I will tell you that, in fact, the figure portrays an old lady with black bangs and an elongated nose. It might take a little while for your perception to change from that of a beautiful girl to that of an elderly lady, but after a moment or two the old lady will emerge.

Initially, you had expected the figure to depict a well-dressed mademoiselle. When examining the illustration, you were actively searching for clues that would conform to your expectations. Eventually, your brain detected those clues and interpreted the figure as a portrait of a young woman. I then revealed that the drawing was not of a beautiful young woman after all, but of a sad-looking old lady. You quickly updated your expectations and searched for the old woman in the figure until, magically, perception matched prediction. In truth, the figure portrays both an elderly woman and an attractive girl. Whether you see one or the other is determined by what you expect to perceive.

The optimism bias relies on a similar principle to turn prediction into reality. First, it alters *subjective* reality. In other words, optimistic beliefs modify our view of the people and events we come across. Most of life's incidents are composed of positive and negative elements. Imagine, for example, that you are a recent graduate of Le Cordon Bleu. You are about to take a new job as the head chef of one of Mario Batali's Italian restaurants in Manhattan—Babbo. You have always dreamed of working alongside Mario's famed red ponytail. The job is prestigious and pays well. However, the package includes long working hours, lots of onion chopping, and a less-than-desirable commute from your apartment in Queens. All in all, do you anticipate enjoying your new job?

After reading chapter 5, you should know that (a) this is a trick question, one that you cannot answer reliably, and (b) long commutes are really terrible for your well-being. Nevertheless, I suspect that dopaminergic neurons fire away when you imagine yourself in a toque, just as Tim's did when he thought of his upcoming trip to Costa Rica. Most of you will readily admit that, yes, you anticipate chopping away with a grin on your face—images of crowded subway trains do not even enter your mind.

As a general rule, we expect the future to be bright. We are, therefore, at least slightly biased to perceive the positive more clearly than the negative. While I am writing these lines, a friend of mine calls. He is at Heathrow Airport, waiting to get on a plane to Austria for a skiing holiday. His plane has been delayed for three hours already due to snowstorms at his destination. "I guess this is both a good and a bad thing," he says. Waiting at the airport is not pleasant, but his mind quickly concludes that snow today means better skiing conditions tomorrow. The possibility that his flight will be canceled and that he will not be sliding happily down the slopes the next day has not yet occurred to him. Eventually, the flight is indeed canceled. Twenty-four hours later, however, he arrives at his destination. The sun is shining and the snow is plentiful.

A canceled flight is hardly a tragic event, but even when the incidents that befall us are the type of horrific events we never expected to encounter, we automatically seek evidence confirming that our misfortune is a blessing in disguise.[11] This is why when most of us think of cancer, we see an old lady, while Lance Armstrong sees a mademoiselle in feathers. No, we did not anticipate losing a job, being ill, or getting a divorce. But when those incidents occur, we search for the upside. We think these experiences help us to mature. We believe they may lead to more fulfilling jobs and stable relationships in the future. Interpreting a misfortune as a pretty young girl means we can maintain that we were correct in assuming things would work out for the best.

Recording brain activity while these swift transformations

take place reveals that highlighting the positive within the negative involves, once again, a tête-à-tête between the frontal cortex and the subcortical regions processing emotional value. While we contemplate a mishap, such as a flight delay, activity in the frontal cortex modulates signals in the striatum that convey the good and bad of the event in question—biasing their activity in a positive direction.[12] The new and improved valuation of the delayed flight is then conveyed back to the frontal regions, and we conclude that the delay is not that bad after all.

Ironically, perceiving setbacks as opportunities may just make them so. This is because predictions not only alter perception but also modify action, thereby changing objective reality. Karl Friston, a professor at University College London and one of the leading neuroscientists today, says, "We will constantly alter our relationship with our environment so that our expectations become self-fulfilling prophecies."[13] Consider a simple example. You expect to meet an old friend at a party. This anticipation triggers a certain behavior—you move about the room, looking at the faces around you; you even stand on your tiptoes to get a better view of the crowd. These actions make it more likely that you will indeed encounter your friend. According to Friston, "This principle may encompass our entire navigation of the world to avoid the unexpected."[14]

According to this reasoning, optimists will carry out actions that will make their rosy predictions more likely. Coach Riley made the L.A. Lakers sweat for twelve months because he believed they could win the NBA championship at the end of the year—and they did.[15] Elaine, the optimistic participant on the show *Survivor,* searched the island for coconuts and swam in the ocean, looking for fish—activities that increased her chances of survival. Matt Hampson is building a house and coaching rugby while in a wheelchair because he believes life is worth living even if he is paralyzed—and so it is.[16] Peter avoids cheeseburgers and goes on long walks because he believes he can avoid another cardiac arrest, which, in turn, makes another incident less likely.

Riley was not the only NBA coach who thought his team would win the 1988 championship. Elaine was not the only participant who expected to be the last survivor standing. Most people believe they have a shot at the prize. Most return empty-handed—this is the essence of the optimism bias. However, those who do not anticipate holding the championship cup, living a healthy life, or achieving professional goals are less likely to carry out actions that can lead to those objectives.

It is tempting to speculate that optimism was selected for during evolution precisely because positive expectations enhance the probability of survival. The fact that optimists live longer and are healthier,[17] combined with statistics indicating that most humans display optimistic biases,[18] together with recent data linking optimism to specific genes,[19] strongly supports this hypothesis.

Yet, with all the good that optimistic illusions have to offer, there are potential pitfalls. There is a danger that in certain situations the relatively small biases of different individuals will combine to create a much larger illusion, which can lead to disaster. Take, for example, the credit crunch of 2008. Each investor, homeowner, banker, and economic regulator expected slightly better profits than were realistically warranted.[20] On its own, each bias would not have created huge losses. However, when combined all together in one market, you got a giant financial bubble, which, when it burst, generated large losses for many individuals.

Consider another example discussed in chapter 11—the construction of the Sydney Opera House, which took a decade longer than initially expected. The faulty project planning was not due to one hugely optimistic individual. In determining a completion date, the project manager needed to take into account estimates provided by several team members—the construction manager, the design engineer, the construction engineer, and the project architect. Being human, each one of these professionals slightly underestimated the time needed to complete the job. Most tasks are performed in sequence rather than simultaneously. In this case, the architect and his team needed to finish the drawings before the construction team could start its work. The slight underestimations of each member accumulated, resulting in significant delays.

At an individual level, optimism can also lead to unwanted outcomes. This is especially true for extreme optimists, for whom the disadvantages of the optimism bias may outweigh the advantages. However, if we are made aware of the bias, we should be able to remain optimistic—benefiting from the fruits of optimism—while at the same time being able to promote action that will guard us from the pitfalls of unrealistic optimism. This is analogous to perceiving the young mademoiselle in Figure 3 while simultaneously knowing that the old lady exists, too. Just as a pilot is able to rely on a plane's navigational system, even when he feels it is guiding him directly toward the ground, we should be able to believe we will live a long, healthy life but also go for frequent medical screening, be certain that our marriage will last but also sign a prenuptial agreement, estimate completing a project in seven months but then add another month to that estimate.

Pilots do not enter the world with an understanding of vertigo. If they were not made explicitly aware of this phenomenon before being allowed to step into the cockpit, aircraft would go down in a graveyard spin on a daily basis. Similarly, the mind is not endowed with an innate understanding of its cognitive biases; neither are we naturally aware of the advantages and drawbacks of these biases. These have to be identified by careful observation, then proved by controlled experiments, and finally communicated to the rest of us.

The brain provides a distorted view of reality. It deceives, yes. But it does so for a reason, and at the very same time it allows the realization that each of us is susceptible to illusions and biases.

Acknowledgments

I am extremely fortunate to have kind, talented, and intelligent friends who also happen to be my colleagues. They not only enhanced the quality of this book by reading chapters, making comments, and offering suggestions, but also provided cheerful support, rendering the process more enjoyable. Tamara Shiner patiently read every single word I wrote and listened to all my dilemmas, of which I had many. She took on the role of friend, collaborator, therapist, and physician. I am eternally grateful. Amir Doron, an author of a brilliant series of books for teenagers, helped me navigate through the early stages of *The Optimism Bias.* Amir is a walking, breathing Google search engine who contributed ideas to many of the examples in this book. I am thankful for taking that empty seat beside him many years ago at our first undergraduate economics lecture. The extraordinary Rosalyn Morn provided help with everything from program code to graceful solutions to social and professional conflicts. Sara Bengtsson gave particularly insightful advice after reading this book; her innovative work inspired chapter 3. Ana Stefanovic read the book with a careful eye and pointed out errors. Patrick Freund, a fun companion, read chapters and provided suggestions. Marc Guitart Masip offered comments and engaged in lengthy discussions. Nick Wright brought to my attention relevant news items and provided commentary. Special thanks to Steve Fleming, whose stubborn support throughout this adventure made me optimistically biased. Our frequent interactions keep challenging my thought and improving my science.

I am especially grateful to my mentor, Elizabeth Phelps. Liz is not only a renowned scientist but also an exceptional woman and adviser. I dread to think where I would be had I not stumbled unannounced into her office more than a decade ago. I owe to her my passion for neuroscience and the constant aim of conducting meaningful research. Liz introduced me to Ray Dolan, one of the leading cognitive neuroscientists working today, who was kind enough to take me under his wing. I am thankful to Ray for being a helpful mentor and close collaborator, and for offering me a spot at the Wellcome Trust Centre for Neuroimaging at University College London. Most of the research described in this book was conducted while I was there. One cannot envision a more dynamic and fruitful place. It is truly a unique institution that gathers together the most gifted and thoughtful scientists.

The idea of turning my research on optimism into a book came from Richard T. Kelly. Richard approached Kevin Conroy Scott from Tibor Jones & Associates, who later became my agent. I thank them both deeply for initiating *The Optimism Bias* and helping to make it into a book. Thanks also to Sophie Lambert and Marika Lysandrou from Tibor Jones. I am especially indebted to Dan Frank, my editor at Pantheon Books, who put his faith in *The Optimism Bias* at the exact time the financial markets were collapsing. I was fortunate to have such an insightful editor as Dan, whose experience and calm demeanor gave me great confidence. Thanks also to the patient Jillian Verrillo from Pantheon Books and to my editor at Knopf Canada, Diane Martin, who was optimistic from the get-go.

I owe my students recognition and gratitude for their hard

work. Christoph Korn in particular made a large contribution to the research described in this book, as did Cristina Velasquez, Candace Raio, Alison Riccardi, Arshneel Kochar, Annemarie Brown, David Johnson, Katelyn Gulbransen, and Elizabeth Martorella. I would also like to acknowledge the numerous scientists whose work is described here, especially Karl Friston, Daniel Gilbert, Daniel Kahneman, Eleanor Maguire, Nicky Clayton, and Laurie Santos. Thanks to the British Academy for supporting my research and to my collaborators Benedetto De Martino, Yadin Dudai, Mauricio Delgado, and Andrew Yonelinas.

Finally, a huge thank-you to the other important people in my life, who, with the exception of one, are not neuroscientists: My friends Keren Sarbero Sorek and Maya Margi for their insight and support. My father, who inspired me to become an academic and whose ideas contributed to chapter 4. My mother, from whom I suspect I inherited a deep interest in human nature. My sibling, Dan, who advised me on all nonscience book-related documents, taking on the role of protective older brother (although he is actually younger). My partner, Josh McDermott, for suggesting important edits, keeping me on my toes, and making my life much more pleasurable.

Notes

PROLOGUE

1. T. Sharot et al., "How Personal Experience Modulates the Neural Circuitry of Memories of September 11," *Proceedings of the National Academy of Sciences of the United States of America* 104, no. 1 (2007): 389–94.
2. Daniel L. Schacter and Donna Rose Addis, "Constructive Memory: The Ghosts of Past and Future," *Nature* 445, no. 7123 (2007): 27, doi:10.1038/445027a.
3. Donna Rose Addis, Alana T. Wong, and Daniel L. Schacter, "Remembering the Past and Imagining the Future: Common and Distinct Neural Substrates During Event Construction and Elaboration," *Neuropsychologia* 45, no. 7 (2007): 1363–77, doi:10.1016/j.neuropsychologia.2006.10.016.
4. T. Sharot et al., "Neural Mechanisms Mediating Optimism Bias," *Nature* 450, no. 7166 (2007): 102–5.
5. Mariellen Fischer and Harold Leitenberg, "Optimism and Pessimism in Elementary School-Aged Children," *Child Development* 57, no. 1 (1986): 241–48.
6. Derek M. Isaacowitz, "Correlates of Well-being in Adulthood and Old Age: A Tale of Two Optimisms," *Journal of Research in Personality* 39, no. 2 (2005): 224–44.
7. Neil D. Weinstein, "Unrealistic Optimism About Susceptibility to Health Problems: Conclusions from a Community-Wide Sample," *Journal of Behavioral Medicine* 10, no. 5 (1987): 481–500.
8. Ibid.; N. D. Weinstein, "Unrealistic Optimism About Future Life Events," *Journal of Personality and Social Psychology* 39, no. 5 (1980): 806–20.
9. Weinstein, "Unrealistic Optimism."

10. Gottfried Wilhelm Leibniz, *Essais de Théodicée sur la bonté de Dieu, la liberté de l'homme et l'origine du mal* (Paris, 1710).

CHAPTER 1

1. Documentary TV series *Mayday,* season 4, episode 9: "Vertigo."
2. David Evans, "Safety: Mode Confusion, Timidity Factors," *Avionics Magazine,* July 1, 2005, http://www.avionicstoday.com/av/issue/columns/993.html.
3. Ibid.
4. Ibid.
5. Ibid.
6. U.S. Summary Comments on Draft Final Report of Aircraft Accident Flash Airlines Flight 604, Boeing 737-300, SU-ZCF, www.ntsb.gov/events/2006/flashairlines/343220.pdf.
7. Ibid. The Egyptian investigative team did not reach the same conclusions as the U.S. team.
8. "Kennedy Crash Bodies Recovered," BBC News, July 22, 1999.
9. Student Pilot—Flight Training Online, "Disorientation (Vertigo)," http://www.news/bbc.co.uk./1/hi/world/americas/401243.stm.
10. Ibid.; Eric Nolte, "Heart over Mind: The Death of JFK, Jr.," Airline Safety.com.
11. Student Pilot—Flight Training Online, "Disorientation (Vertigo)."
12. "Kennedy Crash Bodies Recovered."
13. U.S. Summary Comments on Draft Final Report.
14. Documentary TV series *Mayday.*
15. U.S. Summary Comments on Draft Final Report.
16. E. H. Adelson, "Lightness Perception and Lightness Illusions," in *The New Cognitive Neurosciences,* 2d ed., ed. M. Gazzaniga (Cambridge, MA: MIT Press, 2000), pp. 339–51.
17. P. Thompson, "Margaret Thatcher: A New Illusion," *Perception* 9 (1980): 483–84.
18. N. Kanwisher, J. McDermott, and M. M. Chun, "The Fusiform Face Area: A Module in Human Extrastriate Cortex Specialized for Face Perception," *Journal of Neuroscience* 17, no. 11 (1997): 4302–11.
19. Oliver Sacks, *The Man Who Mistook His Wife for a Hat* (1970; reprint, New York: Picador, 1986).
20. Paul Ekman, *Emotions Revealed: Understanding Faces and Feelings* (London: Weidenfeld & Nicolson, 2003).
21. See http://sallyssimilies.blogspot.com/2008/02/boy-george-looks-like-margaret-thatcher.html.
22. G. Rhodes et al., "Expertise and Configural Coding in Face Recognition," *British Journal of Psychology* 80 (1989): 313–31.

23. Ibid.
24. Ikuma Adachi, Dina P. Chou, and Robert R. Hampton, "Thatcher Effect in Monkeys Demonstrates Conservation of Face Perception Across Primates," *Current Biology* 19, no. 15 (2009): 1270–73, doi:10.1016/j.cub.2009.05.067.
25. Mark D. Alicke and Olesya Govorun, "The Better-Than-Average Effect," in *The Self in Social Judgment,* ed. Mark D. Alicke et al. (New York: Psychology Press, 2005), pp. 85–108.
26. O. Swenson, "Are We All Less Risky and More Skillful Than Our Fellow Drivers?" *Acta Psychologica* 47, no. 2 (1981): 145–46, doi:10.1016/0001-6918(81)90005-6.
27. U.S. Summary Comments on Draft Final Report.
28. E. Pronin, D. Y. Lin, and L. Ross, "The Bias Blind Spot: Perceptions of Bias in Self Versus Others," *Personality and Social Psychology Bulletin* 28 (2002): 369–81.
29. From Emily Pronin's presentation at the Project on Law and Mind Sciences (PLMS) Conference, Harvard Law School, March 8, 2008.
30. Dan Collins, "Scalia-Cheney Trip Raises Eyebrows," CBS News, January 17, 2003.
31. Quoted in Dahlia Lithwick, "Sitting Ducks," *Slate,* February 3, 2004.
32. Emily Pronin and M. B. Kugler, "Valuing Thoughts, Ignoring Behavior: The Introspection Illusion as a Source of the Bias Blind Spot," *Journal of Experimental Social Psychology* 43 (2006): 565–78.
33. Timothy D. Wilson, *Strangers to Ourselves: Discovering the Adaptive Unconscious* (Cambridge, MA: Belknap Press, 2002), pp. 159–82.
34. Petter Johansson et al., "Failure to Detect Mismatches Between Intention and Outcome in a Simple Decision Task," *Science* 310, no. 5745 (2005): 116–19, doi:10.1126/science.1111709.
35. Ibid.
36. L. Hall and P. Johansson, "Using Choice Blindness to Study Decision Making and Introspection," in *A Smorgasbord of Cognitive Science,* ed. P. Gärdenfors and A. Wallin (Nora, Sweden: Nya Doxa, 2008), pp. 267–83.
37. Ibid.
38. "How to Make Better Decisions," *Horizon,* BBC, February 2008.
39. T. D. Wilson and J. W. Schooler, "Thinking Too Much: Introspection Can Reduce the Quality of Preferences and Decisions," *Journal of Personality and Social Psychology* 60, no. 2 (1991): 181–92.
40. Loran F. Nordgren and Ap Dijksterhuis, "The Devil Is in the Deliberation: Thinking Too Much Reduces Preference Consistency," *Journal of Consumer Research: An Interdisciplinary Quarterly* 36, no. 1 (2009): 39–46.

CHAPTER 2

1. C. R. Raby et al., "Planning for the Future by Western Scrub-Jays," *Nature* 445, no. 7130 (2007): 919–21, doi:10.1038/nature05575.
2. Virginia Morell et al. "Nicola Clayton Profile: Nicky and the Jays," *Science* 315, no. 5815 (2007): 1074–75.
3. Raby et al., "Planning for the Future by Western Scrub-Jays"; Nicola S. Clayton, Timothy J. Bussey, and Anthony Dickinson, "Can Animals Recall the Past and Plan for the Future?" *Neuroscience* 4, no. 8 (2003): 685–91, doi:10.1038/nrn1180; Sérgio P. C. Correia, Anthony Dickinson, and Nicola S. Clayton, "Western Scrub-Jays Anticipate Future Needs Independently of Their Current Motivational State," *Current Biology* 17, no. 10 (2007): 856–61, doi:10.1016/j.cub.2007.03.063.
4. Endel Tulving, "Episodic Memory: From Mind to Brain," *Annual Review of Psychology* 53 (2002): 1–25, doi:10.1146/annurev.psych.53.100901.135114.
5. Doris Bischof-Köhler, "Zur Phylogenese menschlicher Motivation," in *Emotion und Reflexivität,* ed. Lutz H. Eckensberger et al. (Munich: Urban & Schwarzenberg, 1985), pp. 3–47.
6. Thomas Suddendorf and Michael C. Corballis, "The Evolution of Foresight: What Is Mental Time Travel, and Is It Unique to Humans?" *Behavioral and Brain Sciences* 30, no. 3 (2007): 313–51, doi:10.1017/S0140525X07001975; William A. Roberts, "Mental Time Travel: Animals Anticipate the Future," *Current Biology* 17, no. 11 (2007): R418–20, doi:10.1016/j.cub.2007.04.010.
7. Suddendorf and Corballis, "The Evolution of Foresight."
8. Raby et al., "Planning for the Future by Western Scrub-Jays."
9. Joanna M. Dally, Nathan J. Emery, and Nicola S. Clayton, "Food-Caching Western Scrub-Jays Keep Track of Who Was Watching When," *Science* 312, no. 5780 (2006): 1662–65, doi:10.1126/science.1126539.
10. Correia, Dickinson, and Clayton, "Western Scrub-Jays Anticipate Future Needs."
11. Raby et al., "Planning for the Future by Western Scrub-Jays."
12. Morell et al., "Nicola Clayton Profile."
13. L. R. Bird et al., "Spatial Memory for Food Hidden by Rats (*Rattus norvegicus*) on the Radial Maze: Studies of Memory for Where, What, and When," *Journal of Comparative Psychology* 117 (2003): 176–87.
14. Thomas R. Zentall, "Mental Time Travel in Animals: A Challenging Question," *Behavioral Processes* 72, no. 2 (2006): 173–83, doi:10.1016/j.beproc.2006.01.009.
15. Tammy McKenzie et al., "Can Squirrel Monkeys (*Saimiri sciureus*) Plan for the Future? Studies of Temporal Myopia in Food Choice," *Learning & Behavior* 32, no. 4 (2004): 377–90.
16. Katherine Woollett, Hugo J. Spiers, and Eleanor A. Maguire, "Talent in the Taxi: A Model System for Exploring Expertise," *Philosophical Trans-*

actions of the Royal Society of London B: Biological Sciences 364, no. 1522 (2009): 1407–16, doi:10.1098/rstb.2008.0288.

17. E. A. Maguire et al., "Navigation-Related Structural Change in the Hippocampi of Taxi Drivers," *Proceedings of the National Academy of Sciences of the United States of America* 97, no. 8 (2000): 4398–403, doi:10.1073/pnas.070039597.
18. "Taxi Drivers' Brains 'Grow' on the Job," BBC News, March 14, 2000, http://news.bbc.co.uk/1/hi/677048.stm.
19. Maguire et al., "Navigation-Related Structural Change in the Hippocampi of Taxi Drivers."
20. Ibid.
21. D. W. Lee, L. E. Miyasato, and N. S. Clayton, "Neurobiological Bases of Spatial Learning in the Natural Environment: Neurogenesis and Growth in the Avian and Mammalian Hippocampus," *Neuroreport* 9, no. 7 (1998): R15–27.
22. J. R. Krebs et al., "Hippocampal Specialization of Food-Storing Birds," *Proceedings of the National Academy of Sciences of the United States of America* 86, no. 4 (1989): 1388–92.
23. Lee, Miyasato, and Clayton, "Neurobiological Bases of Spatial Learning."
24. T. V. Smulders, A. D. Sasson, and T. J. DeVoogd, "Seasonal Variation in Hippocampal Volume in a Food-Storing Bird, the Black-Capped Chickadee," *Journal of Neurobiology* 27, no. 1 (1995): 15–25, doi:10.1002/neu.480270103.
25. J. C. Reboreda, N. S. Clayton, and A. Kacelnik, "Species and Sex Differences in Hippocampus Size in Parasitic and Non-Parasitic Cowbirds," *Neuroreport* 7, no. 2 (1996): 505–8.
26. L. F. Jacobs et al., "Evolution of Spatial Cognition: Sex-Specific Patterns of Spatial Behavior Predict Hippocampal Size," *Proceedings of the National Academy of Sciences of the United States of America* 87, no. 16 (1990): 6349–52.
27. Tulving, "Episodic Memory."
28. Demis Hassabis et al., "Patients with Hippocampal Amnesia Cannot Imagine New Experiences," *Proceedings of the National Academy of Sciences of the United States of America* 104, no. 5 (2007): 1726–31, doi:10.1073/pnas.0610561104.
29. Donna Rose Addis, Alana T. Wong, and Daniel L. Schacter, "Remembering the Past and Imagining the Future: Common and Distinct Neural Substrates During Event Construction and Elaboration," *Neuropsychologia* 45, no. 7 (2007): 1363–77, doi:10.1016/j.neuropsychologia.2006.10.016.
30. Stephanie M. Matheson, Lucy Asher, and Melissa Bateson, "Larger, Enriched Cages Are Associated with 'Optimistic' Response Biases in Captive European Starlings (*Sturnus vulgaris*)," *Applied Animal Behaviour Science* 109 (2008): 374–83.

31. Ajit Varki, "Human Uniqueness and the Denial of Death," *Nature* 460, no. 7256 (2009): 684, doi:10.1038/460684c.
32. Ibid.

CHAPTER 3

1. Lyle Spencer, "Walking the Talk," *NBA Encyclopedia, Playoff Edition,* http://www.nba.com/encyclopedia/coaches/pat_riley_1987-88.html.
2. Jack McCallum, "The Dread R Word," *Sports Illustrated,* April 18, 1988, http://sportsillustrated.cnn.com/vault/article/magazine/MAG1067216/4/index.htm.
3. Ibid.
4. Robert K. Merton, *Social Theory and Social Structure,* rev. ed. (New York: Free Press, 1968), p. 477.
5. "Berlin's Wonderful Horse: He Can Do Almost Everything but Talk," *New York Times,* September 4, 1904.
6. Ibid.
7. "'Clever Hans' Again: Expert Commission Decides That the Horse Actually Reasons," *New York Times,* October 2, 1904.
8. Robert Rosenthal and Lenore Jacobson, *Pygmalion in the Classroom,* rev. ed. (New York: Irvington Publishers, 1992).
9. Susan C. Duncan et al., "Adolescent Alcohol Use Development and Young Adult Outcomes," *Drug and Alcohol Dependence* 49, no. 1 (1997): 39–48.
10. T. L. Good, "Two Decades of Research on Teacher Expectations: Findings and Future Directions," *Journal of Teacher Education* (1987): 32–47.
11. Sara L. Bengtsson, Hakwan C. Lau, and Richard E. Passingham, "Motivation to Do Well Enhances Responses to Errors and Self-Monitoring," *Cerebral Cortex* 19, no. 4 (2009): 797–804.
12. M. R. Cadinu et al., "Why Do Women Underperform Under Stereotype Threat? Evidence for the Role of Negative Thinking," *Psychological Science* 16, no. 7 (2005): 572–78.
13. C. M. Steele and J. Aronson, "Stereotype Threat and the Intellectual Test Performance of African Americans," *Journal of Personality and Social Psychology* 69, no. 5 (1995): 797–811.
14. Richard B. Buxton, *Introduction to Functional Magnetic Resonance Imaging: Principles and Techniques,* 2d ed. (New York: Cambridge University Press, 2009), pp. ix–x.
15. Bengtsson, Lau, and Passingham, "Motivation to Do Well Enhances Responses to Errors and Self-Monitoring."
16. Michael S. Gazzaniga, ed., *The New Cognitive Neurosciences,* 2d ed. (Cambridge, MA: MIT Press, 1999), pp. 7–22.
17. R. Saxe, S. Carey, and N. Kanwisher, "Understanding Other Minds:

Linking Developmental Psychology and Functional Neuroimaging," *Annual Review of Psychology* 55 (2004): 87–124.

18. Gazzaniga, ed., *The New Cognitive Neurosciences.*
19. C. S. Carter, M. M. Bostvinick, and J. D. Cohen, "The Contribution of the Anterior Cingulate Cortex to Executive Processes in Cognition," *Reviews in the Neurosciences* 10, no. 1 (1999): 49–57.
20. Ibid.
21. Jonathon D. Brown and Margaret A. Marshall, "Great Expectations: Optimism and Pessimism in Achievement Settings," in *Optimism and Pessimism: Implications for Theory, Research, and Practice,* ed. Edward C. Chang (Washington, D.C.: American Psychological Association, 2000), pp. 239–56.
22. Christopher Peterson and Lisa M. Bossio, "Optimism and Physical Well-being," in Chang, ed., *Optimism and Pessimism,* pp. 126–46.
23. Ibid.
24. Michael F. Scheier, Charles S. Carver, and Michael W. Bridges, "Optimism, Pessimism, and Psychological Well-being," in Chang, ed., *Optimism and Pessimism,* pp. 189–216.
25. Ibid.
26. Peterson and Bossio, "Optimism and Physical Well-being."
27. Manju Puri and David T. Robinson, "Optimism and Economic Choice," *Journal of Financial Economics* 86, no. 1 (2007): 71–99.
28. Ibid.
29. Judi Ketteler, "5 Money Rules for Optimists," CBS MoneyWatch .com, August 18, 2010, http://moneywatch.bnet.com/investing/article/5-money-rules-for-optimists/457670/.

CHAPTER 4

1. David Gardner, "Obama Can Save Us! Polls Show Wave of Optimism Sweeping the Nation," *Daily Mail,* January 17, 2009, http://www.dailymail.co.uk/news/worldnews/article-1119783/Obama-save-says-America-polls-wave-optimism-sweeping-nation.html.
2. Barack Obama, *The Audacity of Hope: Thoughts on Reclaiming the American Dream* (New York: Crown, 2006).
3. Gardner, "Obama Can Save Us!"
4. Ibid.
5. Gallup poll, *USA Today,* January 4 and January 9, 2001.
6. Royal Society of Arts symposium, "Private Optimism vs. Public Despair: What Do Opinion Polls Tell Us?" The symposium, held on November 6, 2008, was organized by Matthew Taylor and included talks by Ben Page, Daniel Finkelstein, Deborah Mattinson, Matthew Taylor, and Paul Dolan. The theme of this chapter was inspired by this event.

7. Barack Obama's inaugural address, January 20, 2009.
8. Barack Obama's victory speech, November 4, 2008.
9. Diana Zlomislic, "New Emotion Dubbed 'Elevation,'" *Toronto Star,* December 11, 2008.
10. Jennifer A. Silvers and Jonathan Haidt, "Moral Elevation Can Induce Nursing," *Emotion* 8, no. 2 (2008): 291–95, doi:10.1037/1528-3542.8.2.291.
11. Gregor Domes et al., "Oxytocin Attenuates Amygdala Responses to Emotional Faces Regardless of Valence," *Biological Psychiatry* 62, no. 10 (2007): 1187–90, doi:10.1016/j.biopsych.2007.03.025.
12. Michael Kosfeld et al., "Oxytocin Increases Trust in Humans," *Nature* 435, no. 7042 (2005): 673–76, doi:10.1038/nature03701.
13. "Overproduction of Goods, Unequal Distribution of Wealth, High Unemployment, and Massive Poverty," memo from President's Economic Council to President Franklin Roosevelt, March 10, 1933, http://amhist.ist.unomaha.edu.
14. Barack Obama's inaugural address, January 20, 2009.
15. See http://www.kennedy-center.org.
16. Gardner, "Obama Can Save Us!"
17. Gallup poll, *USA Today.*
18. Ipsos MORI 2008 Political Monitor, http://www.ipsos-mori.com.
19. BBC poll, January 20, 2009, http://www.globescan.com/news_archives/bbc-obama.
20. Ipsos MORI 2008 Political Monitor.
21. Ibid.
22. Http://en.wikipedia.org/wiki/List_of_countries_by_intentional_homicide_rate.
23. T. Sharot, C. Korn, and R. Dolan, "How Optimism Is Maintained in the Face of Reality," forthcoming.
24. Royal Society of Arts symposium.

CHAPTER 5

1. Ipsos MORI survey, September 2007, http://www.ipsos-mori.com/assets/docs/news/ben-page-the-state-were-in-ascl-conference-2010.pdf.
2. A. Dravigne, "The Effect of Live Plants and Window Views of Green Spaces on Employee Perceptions of Job Satisfaction" (master's thesis, Texas State University, San Marcos, 2006).
3. Ipsos MORI survey.
4. Daniel Kahneman et al., "A Survey Method for Characterizing Daily Life Experience: The Day Reconstruction Method," *Science* 306, no. 5702 (2004): 1776–80, doi:10.1126/science.1103572.
5. Daniel Gilbert, "Does Fatherhood Make You Happy?" *Time,* June 11, 2006.

6. Richard E. Lucas et al., "Reexamining Adaptation and the Set Point Model of Happiness: Reactions to Changes in Marital Status," *Journal of Personality and Social Psychology* 84, no. 3 (2003): 527–39.
7. "Are We Happy Yet?," Pew Research Center, February 13, 2006, http://pewresearch.org/pubs/301/are-we-happy-yet.
8. Daniel Kahneman et al., "Would You Be Happier If You Were Richer? A Focusing Illusion," *Science* 312, no. 5782 (2006): 1908–10, doi:10.1126/science.1129688.
9. R. Layard, *Happiness: Lessons from a New Science* (London: Penguin, 2005), pp. 41–54.
10. P. Brickman, D. Coates, and R. Janoff-Bulman, "Lottery Winners and Accident Victims: Is Happiness Relative? *Journal of Personality and Social Psychology* 36, no. 8 (1978): 917–27.
11. E. Diener and R. Biswas-Diener, "Will Money Increase Subjective Well-being?" *Social Indicators Research* 57 (2002): 119–69.
12. P. Schnall et al., "A Longitudinal Study of Job Strain and Ambulatory Blood Pressure: Results from a Three-Year Follow-up," *Psychosomatic Medicine* 60 (1998): 697–706.
13. Kahneman et al., "Would You Be Happier If You Were Richer?"
14. Ibid.
15. Paul W. Glimcher, *Decisions, Uncertainty, and the Brain: The Science of Neuroeconomics* (Cambridge, MA: MIT Press, 2004), pp. 189–91.
16. Kahneman et al., "Would You Be Happier If You Were Richer?"
17. A. P. Yonelinas, "Components of Episodic Memory: The Contribution of Recollection and Familiarity," *Philosophical Transactions of the Royal Society of London B: Biological Sciences* 356, no. 1413 (2001): 1363–74, doi:10.1098/rstb.2001.0939.
18. E. A. Phelps and T. Sharot, "How (and Why) Emotion Enhances the Subjective Sense of Recollection," *Current Directions in Psychological Science* 17, no. 2 (2008): 147–52.
19. Tali Sharot and Andrew P. Yonelinas, "Differential Time-Dependent Effects of Emotion on Recollective Experience and Memory for Contextual Information," *Cognition* 106, no. 1 (2008): 538–47, doi:10.1016/j.cognition.2007.03.002.
20. F. Fujita and E. Diener, "Life Satisfaction Set Point: Stability and Change," *Journal of Personality and Social Psychology* 88 (2005): 158–64.
21. E. Diener, M. Diener, and C. Diener, "Factors Predicting the Subjective Well-being of Nations," *Journal of Personality and Social Psychology* 69 (1995): 851–64; "The World in 2005: The Economist Intelligence Unit's Quality-of-Life Index," http://www.economist.com/media/pdf/quality_of_life.pdf.
22. M. E. P. Seligman et al., "Positive Psychology Progress: Empirical Validation of Interventions," *American Psychologist* 60 (2005): 410–21.

23. 2005 data from the European Values Study Group & World Values Survey Association, http://www.wvsevsdb.com.
24. A. Campbell, P. E. Converse, and W. L. Rodgers, *The Quality of American Life: Perceptions, Evaluations, and Satisfactions* (New York: Russell Sage Foundation, 1976), pp. 135–69.
25. T. Sharot et al., "Neural Mechanisms Mediating Optimism Bias," *Nature* 450, no. 7166 (2007): 102–5.
26. Ibid.
27. J. M. Williams et al., "The Specificity of Autobiographical Memory and Imageability of the Future," *Memory and Cognition* 24 (1996): 116–25.
28. W. C. Drevets et al., "Subgenual Prefrontal Cortex Abnormalities in Mood Disorders," *Nature* 386, no. 6627 (1997): 824–27.
29. L. B. Alloy and L. Y. Abramson, "Judgment of Contingency in Depressed and Nondepressed Students: Sadder but Wiser?" *Journal of Experimental Psychology: General* 108 (1979): 441–85.

CHAPTER 6

1. American Psychiatric Association, *Diagnostic and Statistical Manual of Mental Disorders,* 4th ed. (Washington, D.C.: American Psychiatric Publishing, 1994).
2. P. W. Andrews and J. A. Thomson, Jr., "The Bright Side of Being Blue: Depression as an Adaptation for Analyzing Complex Problems," *Psychological Review* 116, no. 3 (2009): 620–54.
3. S. Moussavi et al., "Depression, Chronic Diseases, and Decrements in Health: Results from the World Health Surveys," *Lancet* 370 (2007): 851–58.
4. L. Y. Abramson, M. E. Seligman, and J. D. Teasdale, "Learned Helplessness in Humans: Critique and Reformulation," *Journal of Abnormal Psychology* 87, no. 1 (1978): 49–74.
5. M. E. P. Seligman, *Learned Optimism: How to Change Your Mind and Your Life* (New York: Vintage Books, 2006), pp. 3–16.
6. Christopher Peterson, Steven F. Maier, and Martín E. P. Seligman, *Learned Helplessness: A Theory for the Age of Personal Control* (New York: Oxford University Press, 1995), pp. 182–223.
7. Martin E. Seligman, Steven F. Maier, and James H. Geer, "Alleviation of Learned Helplessness in the Dog," *Journal of Abnormal Psychology* 73, no. 3 (1968): 256–62.
8. Peterson, Maier, and Seligman, *Learned Helplessness,* pp. 182–223.
9. Ibid.
10. G. M. Buchanan, C. A. R. Gardenswartz, and M. E. P Seligman, "Physical Health Following a Cognitive-Behavioral Intervention," *Prevention and Treatment* 2, no. 10 (1999), http://www.ppc.sas.upenn.edu/healthbuchanan1999.pdf.

11. M. Olfson and S. C. Marcus, "National Patterns in Antidepressant Medication Treatment," *Archives of General Psychiatry* 66, no. 8 (2009): 848.
12. Catherine J. Harmer, "Serotonin and Emotional Processing: Does It Help Explain Antidepressant Drug Action?" *Neuropharmacology* 55, no. 6 (2008): 1023–28.
13. A. T. Beck et al., *Cognitive Therapy of Depression* (New York: Guilford Press, 1979), pp. 117–66.
14. Harmer, "Serotonin and Emotional Processing."
15. A. Caspi et al., "Influence of Life Stress on Depression: Moderation by a Polymorphism in the 5-HTT Gene," *Science* 301, no. 5631 (2003): 386.
16. Ibid.
17. D. L. Murphy et al., "Genetic Perspectives on the Serotonin Transporter," *Brain Research Bulletin* 56, no. 5 (2001): 487–94.
18. A. R. Hariri et al., "Serotonin Transporter Genetic Variation and the Response of the Human Amygdala," *Science* 297, no. 5580 (2002): 400; A. Heinz et al., "Amygdala-Prefrontal Coupling Depends on a Genetic Variation of the Serotonin Transporter," *Nature Neuroscience* 8, no. 1 (2004): 20–21; T. Canli et al., "Beyond Affect: A Role for Genetic Variation of the Serotonin Transporter in Neural Activation During a Cognitive Attention Task," *Proceedings of the National Academy of Sciences of the United States of America* 102, no. 34 (2005): 12224.
19. L. Pezawas et al., "5-HTTLPR Polymorphism Impacts Human Cingulate-Amygdala Interactions: A Genetic Susceptibility Mechanism for Depression," *Nature Neuroscience* 8, no. 6 (2005): 828–34.
20. H. S. Mayberg et al., "Deep Brain Stimulation for Treatment-Resistant Depression," *Neuron* 45, no. 5 (2005): 651–60.
21. "Gene-Environment Interactions—Seminal Studies (4 of 7)," http://www.youtube.com/watch?v=vLDvhWF3qis&feature=youtube_gdata.
22. Ibid.
23. Ibid.
24. T. Sharot et al., "Neural Mechanisms Mediating Optimism Bias," *Nature* 450, no. 7166 (2007): 102–5.
25. Ibid.
26. J. E. De Neve et al., "Genes, Economics, and Happiness," SSRN eLibrary (February 2010), CES working paper, series no. 2946.
27. G. Tang, unpublished data.
28. E. Fox, A. Ridgewell, and C. Ashwin, "Looking on the Bright Side: Biased Attention and the Human Serotonin Transporter Gene," *Proceedings of the Royal Society B: Biological Sciences* 276, no. 1663 (2009): 1747.

CHAPTER 7

1. "Guinness Comes to Those Who've Waited," http://www.prnewswire.co.uk/cgi/news/release?id=21223.

2. "How to Pour the Perfect Guinness," http://www.esquire.com/the-side/opinion/guinness031207.
3. "Guinness," http://en.wikipedia.org/wiki/Guinness#Pouring_and_serving.
4. "Guinness Comes to Those Who've Waited."
5. G. Loewenstein, "Anticipation and the Valuation of Delayed Consumption," *Economic Journal* 97 (1987), 666–84.
6. M. L. Farber, "Time Perspective and Feeling Tone: A Study in the Perception of Days," *Journal of Psychology* 35 (1953): 253–57.
7. Loewenstein, "Anticipation and the Valuation of Delayed Consumption."
8. Gregory S. Berns et al., "Neurobiological Substrates of Dread," *Science* 312, no. 5774 (2006): 754–58, doi:10.1126/science.1123721.
9. P. C. Fishburn, *Utility Theory for Decision-Making* (New York: Wiley, 1970).
10. S. V. Kasl, S. Gore, and S. Cobb, "The Experience of Losing a Job: Reported Changes in Health, Symptoms and Illness Behavior," *Psychosomatic Medicine* 37, no. 2 (1975): 106–22.
11. Berns et al., "Neurobiological Substrates of Dread."
12. Tali Sharot, Benedetto De Martino, and Raymond J. Dolan, "How Choice Reveals and Shapes Expected Hedonic Outcome," *Journal of Neuroscience* 29, no. 12 (2009): 3760–65, doi:10.1523/JNEUROSCI.4972-08.2009.
13. George Loewenstein, *Choice over Time* (New York: Russell Sage Foundation Publications, 1992).
14. Tali Sharot et al., "Neural Mechanisms Mediating Optimism Bias," *Nature* 450, no. 7166 (2007): 102–5, doi:10.1038/nature06280.
15. Joseph W. Kable and Paul W. Glimcher, "The Neural Correlates of Subjective Value During Intertemporal Choice," *Nature Neuroscience* 10, no. 12 (2007): 1625–33, doi:10.1038/nn2007.
16. P. H. Roelofsma, "Modelling Intertemporal Choices: An Anomaly Approach," *Acta Psychologica* 93 (1996): 5–22.
17. M. Berndsen and J. van der Pligt, "Time Is on My Side: Optimism in Intertemporal Choice," *Acta Psychologica* 108, no. 2 (2001): 173–86.
18. Hal Ersner-Hershfield, G. Elliott Wimmer, and Brian Knutson, "Saving for the Future Self: Neural Measures of Future Self-Continuity Predict Temporal Discounting," *Social Cognitive and Affective Neuroscience* 4, no. 1 (2009): 85–92, doi:10.1093/scan/nsn042.
19. Timothy L. O'Brien, "What Happened to the Fortune Michael Jackson Made?" *New York Times,* May 14, 2006.
20. "U.S. Savings Rate Hits Lowest Level Since 1933," http://www.msnbc.msn.com./id/11098797/ns/business-eye_on_the_economy.
21. O'Brien, "What Happened to the Fortune Michael Jackson Made?"
22. "U.S. Savings Rate Hits Lowest Level Since 1933."

23. Ibid.
24. Richard H. Thaler and Cass R. Sunstein, *Nudge: Improving Decisions About Health, Wealth, and Happiness,* rev. ed. (New York: Penguin, 2009), pp. 105–19.
25. Lisa Marie Presley's MySpace blog, http://blogs.myspace.com.

CHAPTER 8

1. J. W. Brehm, "Post-Decision Changes in the Desirability of Choice Alternatives," *Journal of Abnormal and Social Psychology* 52 (1956): 384–89.
2. L. C. Egan, L. R. Santos, and P. Bloom, "The Origins of Cognitive Dissonance: Evidence from Children and Monkeys," *Psychological Science* 11 (2007): 978–83.
3. M. D. Lieberman et al., "Do Amnesics Exhibit Cognitive Dissonance Reduction? The Role of Explicit Memory and Attention in Attitude Change," *Psychological Science* 2 (2001): 135–40.
4. T. Sharot, B. De Martino, and R. J. Dolan, "How Choice Reveals and Shapes Expected Hedonic Reaction, *Journal of Neuroscience* 29, no. 12 (2009): 3760–65, doi:10.1523/JNEUROSCI.4972-08.2009.
5. M. R. Delgado, "Reward-Related Responses in the Human Striatum," *Annals of the New York Academy of Sciences* 1104 (2007): 70–88.
6. Louisa Egan, Paul Bloom, and Laurie R. Santos, "Choice-Induced Preferences in the Absence of Choice: Evidence from a Blind Two Choice Paradigm with Young Children and Capuchin Monkeys," *Journal of Experimental Social Psychology* 46 (2010): 204–7.
7. T. Sharot, C. M. Velasquez, and R. Dolan, "Do Decisions Shape Preference? Evidence from Blind Choice," *Psychological Science* 21 (2010): 9209–15.
8. "Choosing the Same Partner Over and Over Again: Commitment in a Healthy Marriage," http://www.meridianmagazine.com/LdsMarriage Network/060714same.html.
9. Leon Festinger, *Conflict, Decision and Dissonance* (Palo Alto, CA: Stanford University Press, 1964).
10. D. J. Bem, "Self-Perception: An Alternative Interpretation of Cognitive Dissonance Phenomena," *Psychological Review* 74 (1967): 183–200.
11. J. Cooper, M. P. Zanna, and P. A. Taves, "Arousal as a Necessary Condition for Attitude Change Following Induced Compliance," *Journal of Personality and Social Psychology* 36, no. 10 (1978): 1101–6.
12. T. Sharot et al., "Dopamine Enhances Expectation of Pleasure in Humans," *Current Biology* 19, no. 24 (2009): 2077–80, doi:10.1016/j.cub.2009.10.025.

CHAPTER 9

1. Jim Bishop, *The Day Lincoln Was Shot* (New York: Gramercy, 1984).
2. F. Colgrove, "Individual Memories," *American Psychologist* 10 (1899): 228–55.
3. R. Brown and J. Kulick, "Flashbulb Memories," *Cognition* 5 (1977): 73–99.
4. U. Neisser and N. Harsch, "Phantom Flashbulbs," in *Affect and Accuracy in Recall: Studies of "Flashbulb" Memories,* ed. E. Winograd and U. Neisser (New York: Cambridge University Press, 1992), pp. 9–32.
5. William James, *The Principles of Psychology,* vol. 1 (New York: Henry Holt, 1890), p. 670.
6. J. M. Talarico and D. C. Rubin, "Confidence, Not Consistency, Characterizes Flashbulb Memories," *Psychological Science* 14 (2003): 455–61.
7. T. Sharot et al., "How Personal Experience Modulates the Neural Circuitry of Memories of September 11," *Proceedings of the National Academy of Sciences of the United States of America* 104, no. 1 (2007): 389–94.
8. "Introduction: One Year Later: New Yorkers More Troubled, Washingtonians More on Edge," http://people-press.org/report/160/.
9. H. Klüver and P. C. Bucy, "Preliminary Analysis of Functions of the Temporal Lobes in Monkeys," *Archives of Neurology and Psychiatry* 42 (1939): 979–1000.
10. L. Weiskrantz, "Behavioral Changes Associated with Ablation of the Amygdaloid Complex in Monkeys," *Journal of Comparative and Physiological Psychology* 4 (1956): 381–91.
11. Joseph LeDoux, *The Emotional Brain: The Mysterious Underpinnings of Emotional Life* (London: Phoenix, 1999).
12. Ibid.
13. The example in this context was posted by Ed Yong in "9/11 Memories Reveal How 'Flashbulb Memories' Are Made in the Brain," http://notexactlyrocketscience.wordpress.com/2007/02/25/911-memories-reveal-how-flashbulb-memories-are-made-in-the-brain/.

CHAPTER 10

1. Lance Armstrong and Sally Jenkins, *It's Not About the Bike: My Journey Back to Life* (New York: Berkley Books, 2001), p. 259.
2. P. Brickman, D. Coates, and R. Janoff-Bulman, "Lottery Winners and Accident Victims: Is Happiness Relative?" *Journal of Personality and Social Psychology* 36 (1978): 917–27.
3. Peter A. Ubel, George Loewenstein, and Christopher Jepson, "Disability and Sunshine: Can Hedonic Predictions Be Improved by Drawing Attention to Focusing Illusions or Emotional Adaptation?" *Journal of Experimental Psychology: Applied* 11, no. 2 (2005): 111–23.

4. Ibid.
5. "The Big Interview: Matt Hampson," *Sunday Times* (London), March 12, 2006.
6. T. D. Wilson et al., "When to Fire: Anticipatory Versus Postevent Reconstrual of Uncontrollable Events," *Personality and Social Psychology Bulletin* 30 (2004): 340–51.
7. T. Sharot, T. Shiner, and R. Dolan, "Experience and Choice Shape Expected Aversive Outcomes," *Journal of Neuroscience* 30, no. 27: 9209–15.
8. Elizabeth A. Phelps and Joseph E. LeDoux, "Contributions of the Amygdala to Emotion Processing: From Animal Models to Human Behavior," *Neuron* 48, no. 2 (2005): 175–87, doi:10.1016/j.neuron.2005.09.025.
9. M. A. Changizi and W. G. Hall, "Thirst Modulates a Perception," *Perception* 30 (2001): 1489–97.
10. E. Balcetis and D. Dunning, "Cognitive Dissonance and the Perception of Natural Environments," *Psychological Science* 10 (2007): 917–21.
11. Leon Festinger, Henry W. Riecken, and Stanley Schachter, *When Prophecy Fails* (New York: HarperPerennial, 1964).

CHAPTER 11

1. Leopold Trepper, *Great Game: Story of the Red Orchestra* (London: Sphere, 1979).
2. R. J. Overy, *The Dictators: Hitler's Germany and Stalin's Russia* (New York: W. W. Norton, 2004), pp. 83–90.
3. Trepper, *Great Game.*
4. Ibid.
5. Ibid.
6. Ibid.
7. Statistic from the American Cancer Society, http://www.cancer.org.
8. N. D. Weinstein, "Unrealistic Optimism About Future Life Events," *Journal of Personality and Social Psychology* 39, no. 5 (1980): 806–20.
9. L. Baker and R. Emery, "When Every Relationship Is Above Average: Perceptions and Expectations of Divorce at the Time of Marriage," *Law and Human Behavior* 17 (1993): 439–50.
10. Neil D. Weinstein, "Unrealistic Optimism About Susceptibility to Health Problems: Conclusions from a Community-wide Sample," *Journal of Behavioral Medicine* 10, no. 5 (1987): 481–500.
11. Overy, *The Dictators,* pp. 483–99.
12. Gabriel Gorodetsky, *Grand Delusion: Stalin and the German Invasion of Russia* (New Haven, CT: Yale University Press, 2001), pp. 67–86.
13. Edward E. Ericson, *Feeding the German Eagle: Soviet Economic Aid to Nazi Germany, 1933–1941* (Westport, CT: Greenwood, 1999), p. 162.
14. Richard S. Sutton and Andrew G. Barto, *Reinforcement Learning: An Introduction* (Cambridge, MA: MIT Press, 1998).

15. David Dunning, Chip Heath, and Jerry M. Suls, "Flawed Self-Assessment: Implications for Health, Education, and the Workplace," *Psychological Science in the Public Interest* 5, no. 3 (2004): 69–106.
16. R. Schulz et al., "Pessimism, Age, and Cancer Mortality," *Psychology and Aging* 11, no. 2 (1996): 304–9.
17. M. F. Scheier et al., "Dispositional Optimism and Recovery from Coronary Artery Bypass Surgery: The Beneficial Effects on Physical and Psychological Well-being," *Journal of Personality and Social Psychology* 57, no. 6 (1989): 1024–40.
18. Manju Puri and David T. Robinson, "Optimism and Economic Choice," *Journal of Financial Economics* 86, no. 1 (2007): 71–99.
19. Thomas Gilovich, Dale Griffin, and Daniel Kahneman, *Heuristics and Biases: The Psychology of Intuitive Judgment* (New York: Cambridge University Press, 2002), pp. 250–70.
20. Peter Jones, *Ove Arup: Masterbuilder of the Twentieth Century* (New Haven, CT: Yale University Press, 2006), p. 214.
21. Peter Murray, *The Saga of Sydney Opera House: The Dramatic Story of the Design and Construction of the Icon of Modern Australia* (London: Routledge, 2003), pp. 56–70.
22. Her Majesty's Treasury, Green Book, http://www.hm-treasury.gov.uk/data_greenbook_index.htm.
23. Hersh Shefrin, "How Psychological Pitfalls Generated the Global Financial Crisis," http://ssrn.com/abstract-1523931; Peter Ubel, "Human Nature and the Financial Crisis," *Forbes,* February 22, 2009.

EPILOGUE

1. N. D. Weinstein, "Unrealistic Optimism About Future Life Events," *Journal of Personality and Social Psychology* 39, no. 5 (1980): 806–20; Neil D. Weinstein, "Unrealistic Optimism About Susceptibility to Health Problems: Conclusions from a Community-wide Sample," *Journal of Behavioral Medicine* 10, no. 5 (1987): 481–500.
2. E. Pronin, D. Y. Lin, and L. Ross, "The Bias Blind Spot: Perceptions of Bias in Self Versus Others," *Personality and Social Psychology Bulletin* 28 (2002): 369–81.
3. T. Sharot et al., "Neural Mechanisms Mediating Optimism Bias," *Nature* 450, no. 7166 (2007): 102–5, doi:10.1038/nature06280.
4. Manju Puri and David T. Robinson, "Optimism and Economic Choice," *Journal of Financial Economics* 86, no. 1 (2007): 71–99; Edward C. Chang, ed., *Optimism and Pessimism: Implications for Theory, Research, and Practice* (Washington, D.C.: American Psychological Association, 2000).
5. Michael S. Gazzaniga, ed., *The New Cognitive Neurosciences,* 2d ed. (Cambridge, MA: MIT Press, 1999).

6. Elizabeth A. Phelps and Joseph E. LeDoux, "Contributions of the Amygdala to Emotion Processing: From Animal Models to Human Behavior," *Neuron* 48, no. 2 (2005): 175–87, doi:10.1016/j.neuron.2005.09.025.
7. E. Tulving and H. J. Markowitsch, "Episodic and Declarative Memory: Role of the Hippocampus," *Hippocampus* 8, no. 3 (1998): 198–220.
8. M. R. Delgado, "Reward-Related Responses in the Human Striatum," *Annals of the New York Academy of Sciences* 1104 (2007): 70–88.
9. K. Friston, "The Prophetic Brain," *Seed,* January 27, 2009, http://seedmagazine.com/content/article/the_prophetic_brain/P1.
10. Pronin, Lin, and Ross, "The Bias Blind Spot."
11. Daniel Gilbert, *Stumbling on Happiness* (New York: Vintage, 2007).
12. T. Sharot, T. Shiner and R. Dolan, "Experience and Choice Shape Expected Aversive Outcomes," *Journal of Neuroscience* 30, no. 27 (2010): 9209–15.
13. Friston, "The Prophetic Brain."
14. Ibid.
15. Mark Heisler, *The Lives of Riley* (New York: Macmillan, 1994).
16. "The Big Interview: Matt Hampson," *Sunday Times* (London), March 12, 2006.
17. Chang, ed., *Optimism and Pessimism.*
18. Weinstein, "Unrealistic Optimism About Susceptibility to Health Problems."
19. E. Fox, A. Ridgewell, and C. Ashwin, "Looking on the Bright Side: Biased Attention and the Human Serotonin Transporter Gene," *Proceedings of the Royal Society B: Biological Sciences* 276, no. 1663 (2009): 1747–51.
20. Peter Ubel, "Human Nature and the Financial Crisis," *Forbes,* February 22, 2009.

Index

Pages numbers in *italics* refer to illustrations.